They are Here

Compelling Evidence of
Extraterrestrial Ships Present on Earth

They are Here

Compelling Evidence of
Extraterrestrial Ships Present on Earth

by

Francisco Villate and Christopher Lock HonFSAI

Published by the authors
2020

Copyright © 2020 by Francisco Villate and Christopher Lock HonFSAI.
US Library of Congress Copyright © TXu 2-187-158

All rights reserved. This book or any portion thereof may not be reproduced or used in any manner whatsoever without the express written permission of the publisher except for the use of brief quotations in a book review or scholarly journal.

Third revised Edition: 2 November, 2020.

Cover: Billy Meier's Hasenbol UFO picture #164, of 29 March 1976.

ISBN-13: 978-1777155001

"Billy" Eduard Albert Meier and FIGU hold all copyrights to Billy Meier's pictures. Their usage in this book is under permission from Billy Meier for this independent investigation. In some cases, the authors received electronic copies of his negatives.

Distributed by Ingram

Based on previous investigations by the same authors, expanded and complemented with new evidence:

Zahi, Rhal. *Analysis of the Wedding Cake UFO*. US Library of Congress Copyright © TXu 1-875-255

Zahi, Rhal. *An Investigation into the Pendulum UFO*. US Library of Congress Copyright © TXu 1-911-482

Zahi, Rhal and Chris Lock. *An Investigative Analysis of Billy Meier's Energy Ship UFO photos*. US Library of Congress Copyright © TX 7-956-476

Zahi, Rhal and Christopher Lock. *Researching a Real UFO: A Practical Guide to WCUFO Experimentation for Young Scientists*. US Library of Congress Copyright © TXu 2-037-764

Related video links:

www.youtube.com/watch?v=6WHqBvBZOqg (A real ET space ship? (WCUFO))
www.youtube.com/watch?v=JoZKwqptZ2Y (Yes... it is a real UFO (WCUFO))
www.youtube.com/watch?v=_pjcbF1oK8Q (New revelations of an UFO - WCUFO)
www.youtube.com/watch?v=8BI-pt3L9I4 (The Pendulum UFO – Jumping in the Space)
www.youtube.com/watch?v=IKeutVKFbG0 (Dancing UFO – OVNI Danzante)
www.youtube.com/watch?v=YsPK_XvfqeI (Worldwide UFO Controversy)

Publishers' information on www.rhalzahi.com

"Things are not always what they seem; the first appearance deceives many; the intelligence of a few perceives what has been carefully hidden."

— Phaedrus

*In fond memory of pioneering investigators
of the Billy Meier case,
Lt. Col. Wendelle Stevens and Prof. James Deardorff.*

Table of Contents

List of Figures and Tables ix
Acknowledgements xix
Foreword xxi
Preface xxix

Introduction: The Worldwide UFO Controversy 1

Part I: The Wedding Cake UFO - WCUFO 13

Introduction to the WCUFO 15
Billy Meier's Courtyard WCUFO: A Detailed Analysis 27
    Introduction and key findings 27
    Photo #800 analysis 30
    Remarkable details in photo #808 43
WCUFO among the Treetops: An Analysis 59
    Introduction and key findings 59
    Photo #838 analysis 61
    WCUFO: computer model simulation 62
    Sphere reflections: analysis of photos #838 and #834 66
Daylight Full-View Full-Size WCUFO 71
    Big tree or little tree, with proximal WCUFO? 71
Below the WCUFO with Trailer and Generator 83
    Previous Research and Meier's Camera 84
    Which camera Meier used and the WCUFO size 89
    Conclusion 97
    Camera formula confirmation 98
    Features of photo #829 104
WCUFO at Night Analysis 113
    Introduction and key findings 113
    The extraordinary photo #873 114
    A wine-red plasma sheath of ionised air molecules? 116
    The size of this WCUFO 118
    WCUFO location at the SSSC 120
    UFO model suggestion untenable 121
    WCUFO vertical extension 125

      Does the WCUFO *emit* or *reflect* light? 126
      Why is the WCUFO a golden orange colour? 137
      Do WCUFO spheres change their shapes? 139
   Summary of WCUFO: Analysis and Conclusions 141

Part II: An Investigation into *The Pendulum UFO* 145

   Introduction 147
      Changes from the 2014 investigation: 149
   Evidence from Four Videos 151
      Introduction 151
      History and evolution of the film 153
      The four videos and sources used in this investigation 157
      PAL and NTSC formats in *The Pendulum UFO* film 163
   The Beamship Demonstration 169
      Introduction 169
      The demonstration sequence 170
      Intriguing details 172
      1- Forced pendulum 173
      2- A big tree and UFO 197
      3- A smooth sharp turn 211
      4- The "jumps" 213
      5- The sudden UFO flip 220
      6- UFO Wobbling 223
      7- The "pulses" 225
      8- Electrostatic flashes 230
      Conclusions 232
   Practical Pendulum UFO Experiments 235
      Introduction 235
      Finding a computer tool 235
      Experimenting with pendulums 237
      Challenging with a pendulum UFO 238

Part III: The Hasenbol Beamship Demonstration 243

   Encounter with a New Beamship 245
      Introduction 245
      Hasenbol location 247
      The photographic sequence 253

  Available videos 282
No Trick Here 285
  Behind the tree 285
  The "impossible fishing rod" 289
  Beamship rim flashes of light 290
  Unexplained movements 291
  Conclusion: no trick or model 296

Annexes 299
  Annexe A 301
    Billy Meier's Olympus, Ricoh, & movie camera 301
  Annexe B 305
    Disclosure Project witness report 305

Notes 309
References 311

List of Figures and Tables

Figures:

Figure 1 -	Beamship in front of a Norway Spruce tree. Photo #66, 9 July 1975	1
Figure 2 -	Wedding Cake UFO hovering above Meier's front yard. Photo #808 in Meier's album.	4
Figure 3 -	Pendulum UFO. A beamship dancing around a tree.	5
Figure 4 -	Photo #728 from the Meier collection. An Energy Ship is hovering above Meier's parking lot?	9
Figure 5 -	Billy Meier's photo, showing a Plejaren beamship above and a miniscout below sporting three undercarriage spheres.	15
Figure 6 -	Photo #808 shows a 3.5-metre diameter WCUFO with crystals, lenses, golden features and various little parts. Meier's home is behind.	18
Figure 7 -	Wide-angle view of Meier's main property in 1981.	19
Figure 8 -	Eleven photographs of the WCUFO taken by Billy Meier in his front courtyard on 22 October 1980, at 11:23 am.	20
Figure 9 -	Photo #829, 26 March 1981, at 6:19 am. A 3.5 or 7-metre diameter WCUFO.	21
Figure 10 -	Photo #834, 3 April 1981, at 1:10 pm.	21
Figure 11 -	Photo series among the treetops.	22
Figure 12 -	Photo #844, 3 April 1981, at 1:35 pm. A WCUFO casting a massive shadow over a young Norway Spruce.	23
Figure 13 -	Top: Photo #850, 3 April 1981, at 2:33 pm. Bottom: The same photo image processed to enhance and reveal this Norway Spruce tree in front of the WCUFO with its bright red crystals.	24
Figure 14 -	Photo #999, 2 August 1981, at 2:18 am.	25
Figure 15 -	Photo #873, 5 August 1981, at 2:48 am. A classic plasma sheath.	26
Figure 16 -	Photo #800, WCUFO with Meier's house in the background.	30
Figure 17 -	Plan view of Meier's residence and courtyard in 1980. The circle indicates the WCUFO location in photo #800 with its central axis A.	32
Figure 18 -	Details of the reflected dark shapes on photo #800 front bottom-tier spheres.	34
Figure 19 -	Sphere photos taken from the model of the carriage house wall. Left: Sphere photographed at one metre from the photographer reveals a much larger reflection than in Meier's photos. Right: Sphere photographed at around six metres away; the red rectangle inset shows a reflected carriage house image very similar in size to the reflected image in Meier's photos.	35

Figure 20 -	Scale model of the carriage house northeast wall with a rectangular hole to locate an iPhone lens. 35
Figure 21 -	Reflections of the carriage house model on a steel sphere (European numeral system of commas for decimal points). 36
Figure 22 -	Reflecting sphere at 2 m from the camera, taken from a position close to the carriage house wall looking north. Right: Zoomed image. 38
Figure 23 -	Reflecting sphere at 1 m from the camera showing the photographer at the centre of the sphere. Right: Zoomed image. 38
Figure 24 -	Confirming the WCUFO location during Meier's photo session; Erhard Lang holds a 3.5-metre long pole in the SSSC parking lot (the courtyard). 39
Figure 25 -	"Blender" computer-made perspective model of Billy Meier's courtyard 40
Figure 26 -	Rendered WCUFO images at different distances and sizes. 41
Figure 27 -	Billy Meier photo #808 in clear focus displaying remarkable details. 43
Figure 28 -	Photo #808 central sphere details. Right: a polygon approximately delineates the carriage house wall. Delineated at the very centre is the inevitable tiny reflection of the photographer. 44
Figure 29 -	Details of colourful crystals and lenses in Photo #808. 45
Figure 30 -	WCUFO at night. A red-reduced photo #873 reveals the top platform red crystals, and blue crystals on the base looking black. 46
Figure 31 -	WCUFO top tier details showing embossed or engraved stars. 47
Figure 32 -	Details on the little plate at the end of the "V" shaped arcs 48
Figure 33 -	Plastic flower pot tray with embossed stars presented by sceptics as "evidence" that Meier made his WCUFO with readily available materials. 49
Figure 34 -	Drawing of Eduard Albert Meier (Billy Meier) by Remington Robinson and a Ptaah self-portrait. 51
Figure 35 -	WCUFO's movement above Meier's courtyard plotted from his 11 sequential photos. Dots indicate the approximate WCUFO centre in each photo. Meier is proximal to the carriage house wall. The 3.5- metre diameter WCUFO shown assumes a southeast arrival. 56
Figure 36 -	Photo #838 taken at treetop level. The WCUFO is proximal to a tree. 61
Figure 37 -	A computer-generated model of the 7-metre beamship and 3.5-metre WCUFO and the forest in photo #838. 62
Figure 38 -	Sphere reflection on the computer-modelled WCUFO showing the tiny 7-metre beamship image. (European numeral system of commas for decimal points.) 63
Figure 39 -	Reflection on the 3.5-metre WCUFO with a simplified rendition of the surrounding forest. (Assuming a Ricoh camera.) 64

Figure 40 -	Reflection on a 0.55-metre scale model showing nearby trees and the dark rectangle at the sphere's centre representing the photographer and the craft he was on. (Assuming a Ricoh camera.) 65
Figure 41 -	Reflections on actual photo #838 showing a forest around the ship. 66
Figure 42 -	Photo #834. The WCUFO shot through or behind a tree. Sphere images show small, low elevation and therefore distant images of the surrounding forest. 67
Figure 43 -	From photo #834. Enlarged view of a sphere with centrally oriented beamship hidden by the tree. 69
Figure 44 -	Photo #841 (top) shows a distant tree close to the WCUFO. Is this a little tree and tiny UFO model, both far away from the camera? Photo #844 (bottom) tree branch details confirm the WCUFO is in front of a big tree. 72
Figure 45 -	Photo sequence from #840 to #844. 10 April 1981. 74
Figure 46 -	Enhanced photo #844 shows dark green reflections on the WCUFO undercarriage the same dark colour as the shaded green tree. 76
Figure 47 -	Photo #844. Reflections on the left of the WCUFO spheres showing the nearby tree as a dark green triangular shape. 77
Figure 48 -	Photo #844. The outlined polygon shadow cast on the tree branches by the WCUFO that blocks the sun's rays, indicated by the arrows and confirmed by the bright light reflection on the WCUFO front right edge. 78
Figure 49 -	Photo #844. Details of shadows cast on the large tree branches, and reflections on the WCUFO undercarriage. 79
Figure 50 -	Zoom of Photo #841. More details of another WCUFO shadow cast on the sizeable tree and branches together with the WCUFO undercarriage reflection revealing the very close proximity of the two. 80
Figure 51 -	Photo #850. Top: Poster version. Bottom: Enhanced image showing a large WCUFO obscured partly by a large tree in front of it. 81
Figure 52 -	A full-frame view of a WCUFO with trailer & generator. 83
Figure 53 -	Photo #866: Meier's green travel trailer with the treetop piece broken off by a beamship 84
Figure 54 -	Meier's 35 ECR stuck on infinity. (Cameramanuals.org for a manual.) 85
Figure 55 -	Billy Meier's Ricoh 55 mm camera. (Butkus.org) 86
Figure 56 -	Meier's Olympus 35 ECR serial number 200519 ("9" behind the shutter release button). Cropped Photoshop double "contrast" for clarity. 88
Figure 57 -	WCUFO, trailer, and generator locations. Elevation views show possible positions for a 550-mm-sized model proximal to Meier, a 3.5-metre WCUFO, and a 7-metre WCUFO. Top: For the Ricoh 55 mm lens. Bottom: For the Olympus 35 ECR 42 mm lens. 102
Figure 58 -	WCUFO with trailer detail showing tree fronds over the edge of the craft. 104
Figure 59 -	Photo 829 from www.theyfly.com "brightness" enhanced only. 106

Figure 60 -	Convergence on the right side edge of the WCUFO.	108
Figure 61 -	Left-hand side of WCUFO showing the edge of the craft hidden by tree fronds.	109
Figure 62 -	Meier with his Olympus 35 ECR camera in 1979.	111
Figure 63 -	Photo #873. A WCUFO at night.	114
Figure 64 -	Photo #873. An enhanced image of the WCUFO at night showing a wine-red ionised plasma sheath and a terrain below the craft.	115
Figure 65 -	Photo #873. Detail of the wooden pole projecting a single shadow downhill below the WCUFO.	118
Figure 66 -	Photo #873. Photo from old books. Left: The pole below the WCUFO is barely visible inside the circle. Right: The halo evident in old photos and a typical vertical pattern on the halo of low-resolution JPEG files.	119
Figure 67 -	Photo #873 SSSC location. Top: View from the deck. Bottom: View uphill from the footpath towards Meier's home.	120
Figure 68 -	Upward extension of the WCUFO central core.	125
Figure 69 -	Night-time sphere reflecting lights.	126
Figure 70 -	Top: Sphere Template for plotting objects vertically and horizontally. Bottom: Cylinder template for plotting objects horizontally.	128
Figure 71 -	Photo #999 from the book *Photo-Inventarium*.	129
Figure 72 -	Auto detail, photo #999.	131
Figure 73 -	Photo #1000.	132
Figure 74 -	Top: photo #921. Bottom: photo #870. Note the orange plasma sheath around the WCUFO	134
Figure 75 -	Photo #871 details.	136
Figure 76 -	Sphere deformations?	139
Figure 77 -	Model sphere deformations created in Blender. Deformation is an optical illusion caused by reflections on contiguous spheres under high contrast conditions.	140
Figure 78 -	The beamship, tree, and houses show the same level of washed-out bluish-grey tones, while the foreground grass at the bottom reveals darker and more vibrant colours. These picture data tell us this is not a little tree close to a little UFO model at only 17 metres from the camera.	148
Figure 79 -	Types of film deterioration: **1** Dust, **2** vertical scratches, **3** microfibres, **4** emulsion peeling, or falling off the film surface.	152
Figure 80 -	Burned frames in the film.	153
Figure 81 -	Snapshot from the movie *Contact*.	157
Figure 82 -	Snapshot from the video *Hinwil* that was available on the FIGU website.	159
Figure 83 -	Snapshot from the NTSC video in the Nippon TV documentary.	160

Figure 84 - PAL video snapshot from the FIGU documentary "Pendulum UFO – Demonstrationsflüge / Demonstration flights (detail FIGU)". ... 161

Figure 85 - Examples of "first" frames in a film roll. The oval marks might prevail at the beginning of every film roll. A few "bad" splicings of two rolls in the FIGU documentary film produce an occasional rectangular white band at the top of the image, as in the left side image. White marks are also present during the beamship "jumps", but they are different. ... 162

Figure 86 - Interlaced images in PAL format. Two images taken at 50 fps combine into one composed image at 25 fps, showing the *Venetian curtain* effect. ... 164

Figure 87 - Top: Two PAL format interlaced frames of *burning* on the film, produced probably by an electrostatic charge inside the camera. Bottom: An interlaced frame showing *Venetian curtain* lines. ... 165

Figure 88 - Row (a) shows 24 Super 8 film frames in one second. Row (b) shows two interlaced frames at 1/50 that combine in pairs to create the PAL frames in row (c). Row (c) shows frames recorded on a PAL video at 25 fps. ... 166

Figure 89 - Resulting interlaced images or frames showing dust particle combinations from two frames in the original film. Top row: each letter represents a unique dust particle or frame detail. Bottom row: the resulting combination in the PAL or NTSC format... 167

Figure 90 - Sequence of demonstration events. ... 171

Figure 91 - Types of pendulum movements found in Meier's film. The period in each one follows the same physical law dictated by the pendulum length and Earth's gravity. ... 174

Figure 92 - Variations of *the fishing rod model*. ... 180

Figure 93 - The *model fixed from a tree branch above*. ... 184

Figure 94 - Composition of several UFO positions and alleged cords projected from each UFO. None of these lines converges into a single node. ... 186

Figure 95 - An arrangement simulated from actual photos for a *model fixed from a tree branch above*. The model failed to produce the variable pendulum length found in all phases ... 186

Figure 96 - A scaled, *sophisticated model* arrangement to make The Pendulum UFO video. . 191

Figure 97 - The overhead support in the *sophisticated model*. ... 192

Figure 98 - An enhanced image shows Rayleigh and Mie light scattering effects. **A**: The background sky. **B**: The distant house, tree, and beamship in light blue-grey tones. **C**: Image enhancement reveals more vivid colours hidden in the film in the closer foreground than in A or B. ... 197

Figure 99 - Left: A simulation illustrating how a little tree and scale model close to the camera would look. Right: Actual image from Meier's film enhanced by increasing contrast and brightness and no colour input or changes. ... 198

Figure 100 -   Left: Image from the Japanese investigators' film showing the distant house sans the tree. Right: Meier's enhanced film image. The UFO, and tree with no supporting device, seem close to or even behind the house, which is partly visible beneath the tree's lower branches. 200

Figure 101 -   Treetop movement after the UFO passes above it. 202

Figure 102 -   Comparison of the location's background blurriness. 204

Figure 103 -   Left: geometry of a nearby model. Right: geometry of a distant UFO. The angles are identical. 206

Figure 104 -   Different levels of UFO blurriness due to Rayleigh and Mie light scatter. Top left: UFO afar. Bottom left: UFO nearer. Note the differences in size, tone and attenuation. Right: The house, tree, and darker greenery at the bottom. 207

Figure 105 -   Meier's 1976 Hasenbol UFO photo #174, sides and top cropped to show the golden Mie light scatter over its front. 209

Figure 106 -   The difference in image blurriness between the near and far UFO is only explicable in the model theory with a very dense foggy environment. However, in such a foggy evening, the distant house would be invisible. The little model theory fails to explain the difference in blurriness, and the similarity in the darkish grey tone of the house, tree, and UFO present in Meier's video. A closer model must be darker and more vivid still, like the greenery in front of the tree. 210

Figure 107 -   A composite of different images of the UFO performing the smooth sharp turn. Image from Deardorff's website. 211

Figure 108 -   Left: Simple UFO Model. Right: Complex UFO Model with one or two gyroscopes. Meier's film shows a movement similar to the model on the left. 212

Figure 109 -   The three "jump" sequences, from top to bottom, taken from the PAL video *Demonstrationsflüge / Demonstration flights*. 214

Figure 110 -   The first "jump" sequence in four frames from the PAL video. Noticeably, the whole image jumps in the second and third frames with interlaced lines slightly in evidence. 216

Figure 111 -   The second "jump" sequence in four frames from the PAL video. Here the interlaced lines are more evident. 217

Figure 112 -   The third "jump" sequence in four frames from the PAL video. Here also, the interlaced lines are evident. The object in the last frame above the house at the left is a large piece of dust. 218

Figure 113 -   Beamship flip. Left: frames from the PAL video. Right: An added line indicates the beamship's horizontal component, which rotates five degrees in less than 1/24th of a second. 220

Figure 114 -   Left-hand sequence: Normal movement of a model hung from a cord. The model should start a gradual rotation while initiating a lateral movement. Right-hand sequence: The beamship in the film rotates (flips) in less than 1/24th of a second, and then moves to the back. 221

Figure 115 -	A model hung from a cord wobbles around its top support contact point or its geometrical centre, but the beamship wobbling centres explicitly around a spot at the bottom of the beamship.	223
Figure 116 -	First "pulse". Top: The PAL video. Bottom: The Nippon TV documentary.	225
Figure 117 -	The four "pulses" with their corresponding times in the film.	226
Figure 118 -	Fourth "pulse" or "little pulse" detail. The arched line of less intensity also crosses the beamship image.	227
Figure 119 -	The authors' purposely scratched negative showing no accompanying dark line for small or deep scratches, and both types pulled emulsion off the negative. Neither form of scratching produces an accompanying dark line.	228
Figure 120 –	A composition of four examples of internal light charges inside the camera with the times they happen in the film.	230
Figure 121 -	Windows Movie Maker screen.	236
Figure 122 -	Testing with a pendulum.	238
Figure 123 -	Meier's photo #174 of 29 March 1976 at 6:02 pm. The new beamship is in front of the sun to the west and behind a tree.	245
Figure 124 -	Clip from Google Earth. In the upper section, BM marks Billy Meier's encounter location on a hilltop east of Fischenthal town.	247
Figure 125 -	A 1972 aerial photo of the Hasenbol location (top) compared with a 2008 Google Earth image (bottom). **1** marks Meier's location. Trees **2**, **3** and **4** in Meier's photos, along with some other trees, have subsequently disappeared.	249
Figure 126 -	Estimate of due west **W** on photo #174.	250
Figure 127 –	Horizontal terrain elevation (top) and bird's-eye plan view (bottom) for photo #174. Elevation and horizontal scales are in metres. Terrain contours and photo from Google Earth. Number **1** is Meier's location, **2** is the *large tree* with the *small tree* under it, **3** is the beamship, and **4** is the *north tree*. The *south tree* is the large unnumbered orange star. North is to the top.	251
Figure 128 -	Bird's-eye plan view, with the approximate beamship location labelled for each photo. North is the top.	255
Figure 129 -	Photo #151 towards the SSW. (Source FoM.)	257
Figure 130 -	Photo #152 towards the SW. (Source FoM.)	258
Figure 131 -	Photo #153. The beamship has now moved back to the left. (Source FoM.)	259
Figure 132 -	Photo #154. From behind the moped. (Source FoM.)	260
Figure 133 -	Photo #155. Showing beamship towards the south with Meier's Super 8 film camera on a tripod. (Source FoM.)	261
Figure 134 -	Photo #157. Beamship flying very close yet high in the sky. (Source FoM.)	262
Figure 135 -	Photo #168. The beamship again from below (*Photo-Inventarium,* page 85).	263
Figure 136 -	Photo #171. The beamship towards the south. (Source FoM.)	264

Figure 137 - Photo #174. The beamship behind the *large tree* and in front of the sun. 265

Figure 138 - Meier's 29 March 1976 *Hasenbol beamship* photos #174 (left) and #164 (right) with sides and top cropped to show better the atmospheric attenuation and golden Mie light scatter over the front of the beamship. 266

Figure 139 - Tone comparison of beamship, foreground, and horizon: **1** the sky, **2** distant mountain, **3** the dark beamship undercarriage, **4** tree trunk and distant mountain, **5** grass near the tree and **6** grass close to the camera. 268

Figure 140 - Photo taken during the Meier and Frehner Hasenbol hill visit in 1998. The *large tree* is now cut down (rectangular inset). Also missing is the *little tree* close to it towards the south. We see the *south tree* behind and downhill, which is absent from current satellite photos. 269

Figure 141 - Photo #164. The beamship is now slightly closer to Meier, behind the *large tree* and again in front of the sun. (Source: High-resolution jpeg from Christian Frehner.) .. 271

Figure 142 - Top: An early scientific-analytical report on photo #164 proves a forked tree branch is between the ship and the camera, positioning the ship behind the tree (Stevens *UFO Contact... A Preliminary...* page 352). 273

Figure 143 - Top left: Zoom in of Christian Frehner's jpeg of photo #164. Top right: FoM picture. Bottom: Villate's two Photoshop details of Frehner's photo #164. [13] Right: Branches traced over minor details shown in the left picture. 274

Figure 144 - Photo #175. Top: Old Wendelle Stevens photo and an enlargement showing the beamship at a 5% clockwise dip. (Sources FoM.) Bottom: Left, Yaoi / Elders *UFO...Contact from The Pleiades* Japanese version page 44). Right, Tom Welch in 1978 standing by the *large tree* in full leaf with two years further growth. Note the tree branches again over the edge of the craft. 275

Figure 145 - Analysis by Sean Gibbons, H.Dip. Digital Media. Meier's #175 bump-mapped on image manipulation program 'Gimp' (Source Deardorff website). Note: Image tilted 10% anticlockwise compared with the original (see Figure 144). 276

Figure 146 - Photo #176. The beamship now moves closer, back to the SW (Source FoM). 277

Figure 147 - Zoom-in of high-resolution photo #176. The beamship has an estimated 20 dark rectangular ports around its edge, and six windows, two visible here. A "ring" sits on its top. (Source FoM.) 278

Figure 148 - Photo #179. The beamship moves closer to the camera and rotates on its vertical axis now showing the top section from its wide side. (Source FoM.) 279

Figure 149 - Zoomed details of photo #179. Notice the dark lines at the base contours. (Source FoM.). 280

Figure 150 - Photo #181. Full-frame high-resolution beamship photo. (Source FoM.) 281

Figure 151 - Video 1 screenshot. The beamship is towards the south with the camera behind a fir tree in the *north grove*. 282

Figure 152 - Video 2 screenshot. Now the beamship is towards the south-west. Meier zooms with the camera several times. 283

Figure 153 -	Comparison of photo #174 (left) and a snapshot from the Nippon TV film (right). Meier and Yaoi are to the right of the *large tree*. The camera is equidistant from them and the *large tree*.	285
Figure 154 -	*Large tree* height estimation and beamship size had it been precisely the same distance from the camera as the tree.	287
Figure 155 -	Had the beamship been a small 55 cm model, it would be where Yaoi is standing. (Meier is on the right).	287
Figure 156 -	Photo #155 with the estimated necessary fishing rod length of over 15 metres. The human figure is drawn to scale.	289
Figure 157 -	Video 2 flashes of light on the beamship's right side edges.	291
Figure 158 -	Composition of over a dozen UFO positions in Video 2. The UFO is moving left to the right (east to west)	293
Figure 159 -	The wobbling centres at a point close to the base of the beamship. A hanging model must flip or wobble centred close to the cord contact point or in the model's centre, not its base.	295
Figure 160 -	Olympus 35 ECR parts.	301
Figure 161 -	Ricoh SLR camera.	302
Figure 162 -	Nalcom FTL 1000 Synchro Zoom that Meier used.	303

Tables

Table 1 -	Approximate necessary camera distances from the centre of the craft for it to occupy approximately 61% of the photo width.	89
Table 2 -	DoF (Depth of Field) for a 42 mm lens with 35 mm film. Details from DOFMaster 2005 ~ 2019. Dan Fleming. All rights reserved	91
Table 3 -	DoF (Depth of Field) for a 55 mm lens with 35 mm film. Details from DOFMaster 2005 -2019. Dan Fleming. All rights reserved	92
Table 4 -	Distances using the camera formula.	100
Table 5 -	Billy Meier's film phases with pendulum periods and measurements.	177
Table 6 -	Tests 1a and 1b with the "fishing rod".	183
Table 7 -	Test 2 Phase simulations 1 ~ 12 with a nylon cord hung from a flexible tree branch	190
Table 8 -	Distances to a model and a UFO	206
Table 9 -	Camera formula data for Meier's photos studied, with shoot time and camera distance in metres. * marks an estimated value	254

Acknowledgements

We thank Billy Meier for allowing us to investigate some of his photos and videos independently, Professor James Deardorff (R.I.P.) for his valuable comments and challenges to improve these investigations, Christian Frehner for his continuous support and help, and Dyson Devine and Michael Horn for their contributions and support.

We also thank several sceptics and would-be debunkers who caused us to re-think some of our previous research, and whose questions and challenges resulted in further insights.

Foreword

The following comments from Professor Deardorff (R.I.P) concern the initial WCUFO investigation published in the first quarter of 2013.

Commentary by James Deardorff

> For the photo #800 analysis, it's all about how far the WCUFO was from the carriage house (CH) wall, judging from the width of the CH's image in the reflected spheres relative to the diameter of the spheres. For this, Zahi needed a good plan view of precisely where the CH was located with respect to the main residence, which is the building that shows up in the background of photos #800 & 799, which he also analyzed. (So he got that from Google Earth, which didn't yet have a good view of the property when I had done my analysis some 10 years ago.) He used a computer program (called Blender), which accurately places reflected images of surroundings in any mirrors that you designate, even curved and spherical mirrors. So, having the layout of Meier's property and a nice computer model of the WCUFO with its reflecting spheres in proper positions, he could examine the width of the image of the CH in the spheres relative to the diameter of the sphere. Only when he had the model WCUFO at the proper distance from the CH did the reflected image of the CH on the modelled spheres have the same width as in photo #800. This distance was some 20 ft. away.
>
> But he checked this against the reflection from a model sphere (small reflecting sphere; call it a marble) within a physical model of Billy's property, and took photos of that, when the marble was at various locations away from the cardboard-model CH. Only when the marble was the proper distance away did its reflected image of the CH have the proper relative width on the marble sphere. When this distance was scaled up from the model to the real Google Earth dimensions of Billy's property, it corresponded to the WCUFO being some 23 feet away. Each figure has a plus-or-minus error figure, within which they show agreement that

it was this general distance away from the CH – 20 ft. - 23 ft. or so, 6 or 7 meters or so.

But then also, Zahi had to know that the camera & Meier were situated right close to the CH, not 20 feet or so away towards the main residence, using a model. He showed this in four different ways. One way was to show that if Meier & model had been out in the middle of the parking area (courtyard) somewhere, then the main residence in the background would have looked way too large (too close) in the photo. Another way was to show that if Meier had been located with a model much closer to the main residence, he himself would have shown up in the photo because his model would only have been 3 ft. away from the camera. (He proved this by showing a photo of the reflection of someone close to a sphere and also in the computer model.) A third way he showed it: in the reflections of the CH in Meier's hypothetical model, the location of the CH within each sphere of the model would have shifted a bit due to the appreciable distance between the camera and the CH (a parallax effect). Such does not happen within adjacent spheres of the WCUFO of #800 or 799, since Meier's camera was close to the CH wall. The fourth way was by calculating how far away the camera had to be from the main residence, knowing certain dimensions of the main residence and the width it occupied on the film (as well as knowing the camera's focal length and the film size). All four ways agree – Meier had his back close to the CH wall when taking those photos.

But also, Zahi had to know just where Meier was standing relative to the CH wall – was it near one corner or the other or in the middle? But this was obvious to him at first glance of any of the reflections of the spheres on #800 or #799. The cameras location within the reflected image from each sphere is at the exact center of the sphere as you view it. (I had failed to realize this fundamental and obvious fact.) So, Meier had been standing next to the center of the CH's wall.

But then also, to get the dimensions of the CH correct, and the fact that its eaves show up fairly prominently in the reflections, and the fact that the CH underwent some modifications shortly before 1980, he had to check with

Christian pretty closely, and also make use of a frame from the movie "Contact" taken before 1980.

So going from the WCUFO's sphere's distance away from the CH to the WCUFO outer diameter involved a little arithmetic. It was somewhere between 3.0 and 3.6 meters in diameter (Meier and Christian said it was the 3.5m craft). Figure C6 best shows Zahi's result, I think.

His analysis of the treetop WCUFO photos was especially impressive to me. By examining the spheres' reflection closely, in photos #834 & 838, Zahi noticed that the outline of the small forest treetops was evident. This meant that Meier himself was up at treetop level taking the photos (from Quetzal's ship). He showed that if the WCUFO had been a model somehow suspended behind the nearby tree's uppermost branches and in front of the more distant tree, such a model would have had to be very close to that nearby treetop, in which case the latter would have shown up as rather large in the spheres' reflections, which it does not. It also meant that Meier himself (or supposedly a two-armed confederate who had climbed to treetop level) would have shown up in the sphere's image, which of course they do not. Since the WCUFO was not a model, it was farther away than any model would have been – close to the tree behind it – and so Quetzal's ship itself does not show up in the reflections of the spheres in #834 or 838 or others of that series; the forest treetops behind Quetzal's ship did not allow it to stand out against a clear background, and so that craft does not quite show up against the forest treetop background in the reflections from the spheres in #834 or 838.

And then in his 74-page article RZ [Rhal Zahi] included the strong evidence that one of the night time WCUFO shots showed the craft with its upper part extended upwards by a significant amount. This took a bit more geometry for him to show.

Prof. James Deardorff

April 2013

Disclosure Project witness Dyson Devine (Annex B), with Vivienne Legg, has translated into English numerous FIGU publications, Contact Reports on the website futureofmankind.co.uk, pieces for theyflyblog.com, and the "Billy" Eduard Albert Meier books so treasured by many: ***Might of the Thoughts*** and ***The Way to Live.***

Commentary by Dyson Devine

This book, *They are Here* – "they" being the Plejaren Federation spacecraft and their pilots – allows us to draw the personal conclusion that they are not landing at the U.N. or White House because they are very stable geniuses, unlike most but thankfully not all of us here on Earth. Their exotic ET contact methodology lovingly filters out those of us who would be devastated by the actual truth and at the same time powerfully attracts those of us already armed with it. As the spiritual teaching informs, "Truth is powerless against a weak consciousness." Think for yourself.

I thank the authors for the privilege of contributing to their foreword and for the pleasure of our collaboration. *They are Here* represents, among other wonderful things, a microscopically detailed accumulation of indisputable hard physical facts that allows anybody with an average education and intelligence, and just a little common sense, to end the nonsensical "hoax" debate.

Naturally, their book raises endlessly more questions, and some of the answers are as obvious as they are unpalatable. It's human nature to bury unpalatable truths deeply, and it's also the nature of truth that there are never enough shovels to keep it buried, and attempting to results in digging one's own grave as the coming times disclose. But why on Earth must it take so long for otherwise sane and reasonable individuals to accept the increasingly obvious? What might answer this question and not discussed in this book, is the excerpt from the 304th contact of June 2001, which mentions secret, weaponised, portable, insanity-generating technology wielded by the secret agents of the organised crime group Billy calls the Order of Darkness.

Another possible answer could be that according to Ptaah: The Earth human does not want to know the truth.

Aeons of self-censorship, however, have just been quietly penetrated with this piercingly historic book, which, along with the various scientific analyses, also collects various broadly scattered excerpts, from the thousands of pages of translated material, into an undeniable picture, which once seen, cannot be unseen.

In my view, this book is more an explanation of the exotic methodology, as much psychological as technological, used by these benevolent extraterrestrials, which is a message that I think the Plejaren Florena would agree is very important to disseminate. In 2010 she instructed Billy Meier, Core Group members, and others to no longer engage in debates at trying to convince sceptics, and know-alls of the case. Strongly forbidden are evil thoughts of revenge. All the information is now out there for everyone to freely access, and the deniers are too weak in mind and fixed in thinking to recognise the truth. Florena's advice below from the 2010 Contact Report 486 V 6, 12~15, and 17, may also serve well all those who gain insight, knowledge, and understanding from this book:

6. Since the beginning of the existence of our contacts and the release of your evidential materials starting from the year 1975, in this respect, again and again controversy has been triggered by malicious ones, by know-it-alls, critics and calumniators.	6. Seit dem Bestehen unserer Kontakte und der Veröffentlichung deiner Beweismaterialien ab dem Jahr 1975 wurden in dieser Beziehung durch Böswillige, durch Besserwisser, Kritiker und Verleumder immer wieder Kontroversen ausgelöst.
...	...

12. Thereto, according to Ptaah and Quetzal and from all of us who work together with them and discuss all necessary things with them, I now should advise that you and the Core Group members and passive members, as well as Michael Horn and others, in no further way ought to involve yourselves in any controversy in said form.

13. Also, in the correspondence as well as on the FIGU forums and with journalists and visitors, and so forth, the whole thing ought not to be gone into, and thus no questions relevant to controversy or other questions which are directed at calling the truth into question, are to be answered any more.

14. For the longest time, in the same way that evidence of the truth of our collective contacts has already been provided by many witnesses, as has the genuineness of your photos made of our beamships and various materials, which were given by Genesis III in the USA, to renowned scientific experts and institutions for scientific investigation and analysis and which proved

12. Dazu soll ich dir nun gemäss Ptaah und Quetzal und von uns allen, die wir mit ihnen zusammenarbeiten und zusammen auch alle notwendigen Dinge beraten, nahelegen, dass du und die Kerngruppemitglieder sowie die Passivmitglieder, wie aber auch Michael Horn und andere Personen, euch in keiner Weise mehr auf irgendwelche Kontroversen in genannter Form einlassen sollt.

13. Auch in der Korrespondenz sowie in den FIGU-Foren und bei Journalisten und Besuchern usw. soll auf das Ganze nicht mehr eingegangen und folglich auch keine kontroversebetreffende oder sonstige Fragen mehr beantwortet werden, die darauf ausgerichtet sind, die Wahrheit in Frage zu stellen.

14. Die Wahrheit unserer gemeinsamen Kontakte wurde ebenso durch viele Zeugen schon längstens bewiesen, wie auch die Echtheit deiner von unseren Strahlschiffen gemachten Photos und diversen Materialien, die durch die Genesis III in den USA für wissenschaftliche Untersuchungen und Analysen an namhafte wissenschaftliche Fachkräfte und Institute gegeben

that all the material corresponds to genuineness and that no counterfeits whatsoever are presented.	wurden und die bewiesen, dass alles Material der Echtheit entspricht und dass keinerlei Fälschungen vorliegen.
15. Also, the investigations on the sites and locations where the photos, and so forth, were made showed that all your data correspond to the truth and that no deception whatsoever is present.	15. Auch die Abklärungen an Ort und Stelle der Aufnahmen der Photos usw. ergaben, dass alle deine Angaben der Wahrheit entsprechen und dass keinerlei Betrug vorliegt.
...	...
17. Therefore, neither you, nor other members or friends and acquaintances, ought to react to further discriminatory machinations and questions, rather declare that the evidence of the genuineness and truth in regard to the contacts and photos, as well as other materials is sufficiently provided, consequently it is nonsensical to still continue to talk about it.	17. Also sollen weder du noch andere Mitglieder oder Freunde und Bekannte auf weitere diskriminierende Machenschaften und Fragen reagieren, sondern erklären, dass der Beweise der Echtheit und Wahrheit in bezug auf die Kontakte und Photos sowie sonstige Materialien genug gegeben sind, folglich es unsinnig ist, noch weiter darüber zu reden.

The mockers and the know-it-alls will eagerly board the ship of truth only once the passengers start visibly filling her up, and the fear of being left behind alone takes hold. The first to leave the sinking ship of lies will be the rats, whose pernicious presence also partially explains the lengthy time that collective enlightenment is taking us.

By now, anyone who genuinely desires to know that the Billy Meier contacts are legitimate should know, and other sleeping people voluntarily place themselves in a category where they must eventually stumble over the truth by themselves. So don't weaponise this book.

It's also important to remember that one's long lingering doubts are poison to spiritual evolutionary progress and should be eradicated with the logical study of books like this one as soon as practicable.

Finally, as a complement to Florena's words, Billy stresses the inescapable need for *amicable* interpersonal disputation, and how it works to advance our collective consciousness-related evolution, which is our reason for being.

Please accept the truth you find with your understanding and share it justly and lovingly.

Thank you and salome.

Dyson Devine

January 2020

Preface

The Billy Meier case evidence began to accumulate in quantity in the 1970s and 80s when Photoshop, CGI technology, and personal computers were unavailable for faking UFO evidence. Any forgeries created at the time had to use little scale models, multiple exposures, or retouching negatives or prints. Meier's photographic evidence, some of which we investigate in this book, has stimulated the participation of UFO enthusiasts, investigators, sceptics and wannabe debunkers of the case.

Serious UFO investigators come in two types: the 1970s and 80s experts who used sophisticated labs to test the Meier evidence; and present-day investigators performing tests that anyone can replicate to find the truth for themselves. Computer technology available to everyone today provides an excellent means for anyone to perform personal computer analyses that compare well with the expert computer analyses performed in the specialised 1980s laboratories.

Those earlier 1970s and 80s investigations on the Meier case evidence as presented in the Movie *Contact* (Stevens, 1978), were conducted by Wendelle Stevens, Lee Elders, Britt Elders, Tom Welch, Jim Dilettoso, and Marcel Vogel, among others.

More recently, investigators like James Deardorff made analyses on the Pendulum UFO, and *The Wedding-Cake UFOs* – aka WCUFO (Deardorff, 2006). Christian Frehner, following Deardorff's recommendations, made some tests at Billy Meier's property using a reflecting sphere to compare its reflections with those on WCUFO spheres presented in the WCUFO photos.

In 2009 Christopher Lock, coauthor of this book, made initial analyses on the photograph of the WCUFO hovering above a green trailer. This book includes Deardorff's results, and Lock's updated results that go considerably beyond the previously published trailer WCUFO detail in *Researching a Real UFO* (Zahi and Lock page 7). Other curious researchers also have since found that a simple Photoshop function reveals the WCUFO partly behind the tree leaves in this WCUFO photo, showing the craft is a distant object, not a little nearby model used in a false perspective shot.

By the end of 2012 Rhal Zahi (Francisco Villate), coauthor of this book became independently intrigued by a particular WCUFO photo (#800). Like Deardorff and Frehner before him, he noticed some shadowed reflections on its spheres. He then thought to check what was on the premises to gain details that could give a very reasonable estimate of how far away from the camera this WCUFO was and subsequently its size. Inspired by Professor James Deardorff's findings and investigations, he contacted Deardorff and decided to make a more detailed analysis that became his 74-page report: *Analysis of the Wedding Cake UFO* (Zahi 2013). In this report, all the protocols and calculations were described in detail so that anyone with a basic knowledge of junior high school math could replicate them.

Zahi conducted more investigations in 2014, now on *The Pendulum UFO*, and he demonstrated this UFO was a big object dancing around a tree, imitating the movement of an object hung from a cord (pendulum movements). Chris Lock and Dyson Devine helped him refine his findings. Then in 2015 *UFO Digest* published an interesting comment on Zahi's investigation of *The Pendulum UFO* (YouTube video):

> After viewing "El OVNI Danzante" ("The Dancing UFO"), we would like to commend Rhal Zahi and his team on the most detailed and scientific analysis of the Billy Meier "Beamship UFO" that we ever seen [sic] at UFO Digest. We thereby concur with Rhal Zahi's findings that the movement of the Meier "Beamship UFO" as seen in the Meier Film is NOT a small model suspended on a pendulum, orbiting a small fake tree (for the many reason [sic] cited in his film by Rhal Zahi and his team).
>
> These computer graphic and video demonstrations by Zahi & company of various schemes designed to hoax a "Pendulum-Propelled UFO model" in the Billy Meier film are untenable and impracticable, if not impossible for a 3-man team, let alone for a one-armed photographer.
>
> *Robert Morningstar*
> *Editor - UFO Digest (ufodigest.com)*

Also in 2014, we (Zahi and Lock) started our investigations into the Energy Ships, said to be UFOs made of energy displayed in variable forms. Meier and the Plejaren say these strange ships bring to Earth advanced little beings from the Andromeda galaxy. If real, these visitors, perhaps like us just recently, use light as a highly efficient data recording device (Xiongbin). It was evident during our investigation that these ships interacted with the surrounding environment indicating they were not double exposure photographic *tricks* as sceptics had claimed. However, at least one is a triple exposure. Our investigations into these photos show progress, though lacking physical structures there is less to analyse and what there is remains somewhat elusive. We are also grateful to sceptics for pointing out a few errors made in an earlier paper we put out on them and hope to address all of their points in due course.

With more and more evidence analysed over the years and additional hidden clues now revealed, the Meier case evidence is compellingly real.

A classic example is the famous WCUFO photo at night that for approximately three decades had been hiding details within its black background that with brightness enhancement revealed this WCUFO to be hovering above a terrain with a footpath and a visible pole. It manifests no clues suggesting a little model and displays a mysterious pink halo around it that some researchers say is a plasma plume. Estimates made from revealed detail in this night-time picture indicate the WCUFO is approximately seven metres in diameter.

Our recent book *Researching a Real UFO - A Practical Guide to WCUFO Experimentation for Young Scientists* (2017) focused on students performing 11 experiments on the WCUFO evidence to calculate how big these objects were and to discover fascinating details about them. That book shows simple, easy-to-perform tests and presents three useful sphere reflection rules that help readers understand in practical ways how sphere reflections work and the information they provide about this ship.

So, why are we now publishing this book and to what audience? This book is a handy one-volume compilation of three types of Meier evidence: The Wedding Cake UFO, *The Pendulum UFO* film, and a new Hasenbol UFO photographs analysis. Liberally illustrated with nine tables and over 150 colour figures, the book provides a visually-enriched reader experience. It is, we hope, an easy-to-follow and revealing text for any researcher, UFO enthusiast, and anyone interested in the Meier case, UFOs, and the possibility of extraterrestrial lifeforms and their possible visitation to Earth. It also presents a summary of discoveries, and abundant references indicating further investigation details on these UFOs and the Meier case.

With the recent rapid and unparalleled developmental progress in digital and computer technology, investigators of the Meier Case have changed throughout the years. As mentioned, anyone can now make investigations with readily available computer tools without the need to rely on past experts. You, the reader, can now document findings, indicate how to obtain them, and have your conclusions corroborated by anyone with a basic knowledge of junior high school math. Today's computers give an ample, even impressive ability to anyone interested in finding the truth for themselves.

Is the Billy Meier evidence like that discussed in this book the most critical part of this case? Perhaps surprisingly, we think not. In this book, we present proof showing the UFOs are real and that: *They are Here.* This knowledge is an entrance door to access the real treasure inside the case, a far more critical treasure for humanity in this time of social, environmental, and global upheaval and transition: a new philosophy of life, non-religious and belief-free, that Meier calls *the spiritual teaching.*

The spiritual teaching consists of the books and publications of Billy Meier such as *Goblet of the Truth, Arahat Athersata, The Way to Live, The Psyche, OM, Might of the Thoughts* and over 700 detailed individually numbered Contact Reports covering every topic imaginable in face-to-face discussions between Meier and the Plejaren. Currently, these works are most readily accessible in English from FIGU, or Michael Horn's: **http://theyfly.com/** and **http://www.futureofmankind.co.uk/Billy_Meier/Contact_Reports.**

At this pivotal and precarious time in our world's history, the spiritual teaching contains recommendations from Meier and his ET Plejaren friends that can help make the Earth a far better place, and ourselves far better people individually and collectively. Do the UFOs analysed and discussed in this book belong to the Pejaren? We do not *prove* they do, but knowing *They are Here,* what options exist for the identities of the pilots who show their craft exclusively to Meier and other eyewitnesses? No rational explanation exists for these being the product of any government, military, or secret scientific program in Switzerland.

Knowing, as this book shows, that Meier's UFO evidence is real, humanity can now seriously consider whether he has been in contact for several decades with advanced space beings from a place in another space-time configuration beyond the Pleiades star cluster. These ponderings and researches urge us to consider both how these ETs see us on Earth and their recommendations to us in our epochal time.

Meier says the Plejaren have lived in peace for over 50,000 years and have developed an incomprehensibly advanced galaxy and time-travelling technology. Perhaps their words, messages and advice that they and Meier bring to us should be considered. Even if unable to consider their existence factual, at least their recommendations can be read objectively to see what assistance or answers they offer to some of the questions held by most of us working to improve our world. Perhaps chief among those questions is: How can we individually and collectively best survive and evolve to be responsible cosmic inhabitants and ensure our successful destiny as guardians of this delicate, petite, pale blue planet we call our Earth?

We need not, however, for centuries, trouble ourselves to try visiting the Plejaren, assuming we could. Our meeting would probably result in cultural, psychological, and spiritual shocks to both them and us due to our many pandemic primitive traits that they evolved beyond tens of thousands of years ago. These are traits they can neither no longer comprehend nor endure in interpersonal communications and relations.

Meier says the Plejaren have come here to give information that can help us, and to visit him, their friend with whom alone they think and feel capable of communicating. Since childhood, Meier has been writing to help all human beings everywhere throughout the many universes of Creation. Who would not want to hear the advice of people 50,000 years ahead of us?

The spiritual teaching and the Plejaren, however, are not discussed in this book. Those topics are for you, the reader, to move onto if this book resonates with you and you concur with its elementary scientific evaluations and findings. Two excellent places to begin researching the spiritual teaching and the Plejaren are the bold online links on the previous page.

The Plejaren have shown us their crafts, or beamships as they call them. This book provides proof these UFOs are here, so their pilots must be too, either inside them or remotely controlling them.

Who are those pilots if not the Plejaren who over the past several decades have given us their wise words of wisdom, intelligence, knowledge, love, and advice for now and futurity?

Finally, according to Meier, when he dies, the Plejaren visitations to Earth will cease. Meier is now 83 years old, so they may not be here much longer; but right now, for just a little while longer, like their UFOs: *They are Here.*

Francisco Villate (Rhal Zahi) and Christopher Lock

February 2020

Introduction:
The Worldwide UFO Controversy

At first sight, the Meier case presents an extraordinary volume of incredibly compelling evidence. Despite this library of evidence, some say the case is too good to be true. For them, for some reason, initial doubt is created, despite the case's overwhelming proof of the highest quality. Perhaps such uncertainty is, however, understandable.

Thousands of written pages of unique information still await attention and translation into English and, together with accounts of Meier travelling to other galaxies and even into the past and future, there are enough extraordinary details to instil an air of initial scepticism concerning the case. Furthermore, there is no bigger UFO rabbit hole of investigation to burrow into, and such a great challenge results in many people expressing immediate denials of the case's feasibility. Few welcome such a time-consuming investigation of thousands of pages of research, or an upset to their familiar comfort zone.

Figure 1 -   Beamship in front of a Norway Spruce tree. Photo #66, 9 July 1975

They are Here

Many people think that the extraterrestrials contacted Billy Meier and performed their demonstrations to convince us of their existence. We have found that this is not the case. Why, then, show themselves in the way they did? These ETs, or Plejaren as they name themselves, say they made their demonstrations to create a worldwide UFO controversy; a controversy that would result in humanity at large *considering* their possible or likely existence rather than proving it. According to Meier, a Plejaren commander named Ptaah explained in Contact Report #251 (Meier) that they did not attempt to present irrefutable evidence of their presence here on Earth, which, of course, they could have done quite readily. Ptaah's exact explanation follows momentarily when briefly discussing *The Pendulum UFO* demonstration.

Investigators to the Meier case soon face the challenge known as the sceptic barrier. This barrier constitutes claimed evidence against the Meier case, which includes several simple demonstrations and very flimsy research on how Meier supposedly faked his evidence, and unsubstantiated claims of making models that sceptics present as proof. Further complicating investigation is the fact that some Meier fake photos were genuine photos stolen or misappropriated and subsequently falsified by various people to cast doubt on the case. Other sceptics fight against the evidence because it goes against their underlying belief that Meier is a fraud or worse. They also invariably show a lack of investigative tenacity necessary to thoroughly search through or delve into all the possibilities and are often comfortable with making quick, simplistic conclusions. These beliefs and outlooks have resulted in copious defamatory, libellous comments and calumny against Meier's character denigrating him an outright fake and fraud without ever presenting evidence that would stand up in a court of law.

It is all too easy to become blocked or entrapped in this barbaric barricade of sceptic noise, and feel no need or point in proceeding any further to discover whether the case is real or not. So, facing the noisy know-it-all sceptic barrier alone with the Meier evidence is enough for many to discount or drop the case, and it is quite evident sceptics keep buttressing their barrier. What many people are unaware of, however, is that the extraterrestrials knowingly contributed to creating the sceptic barrier and that they had a good reason for doing so.

Introduction: The Worldwide UFO Controversy

In the investigations of the 1970s and 1980s, several tests were made using UFO scale models of around 50 cm in diameter, by Wendelle Stevens and other investigators as presented in the *Contact* movie (Stevens and Elders 1982). The resulting images were very similar to Meier's photos which led to sceptics claiming Meier too used little models. Such "proof" is an example of a first impression judgment, based on initial suppositions to support a hoped-for goal rather than proper, objective, detailed and rigorous research for the whole truth. Christian Frehner recently informed us by email (2020) that Billy Meier says he has never made a model of a beamship. Sceptics have no more than assumptions, claims or made-up stories based on models other people have made to support their idea of Meier making models.

Regarding the photos, for instance, Jim Dilettoso found an unusual characteristic in them that indicates the distance from the camera to the object photographed. In emulsion films, a photographed object against an unobstructed background produces a thin border or line at the edges of the image; the farther away from the object, the wider the border. In this way, he was able to determine approximate distances to the UFOs and differentiate a UFO model from a large flying object. As indicated in *UFO Contact from the Pleiades – A preliminary investigation report* (Stevens *Annex IV* page 380) Dilettoso found no evidence of forgery in Meier's photos by analysing the negatives with sophisticated equipment and procedures. On many occasions Dilettoso has defended the Meier case as real, indicating that the photographs show large real flying objects of around seven metres in diameter. He did so in the 23rd annual National UFO Convention, Phoenix, Arizona, on 16 May 1986 (Stevens *UFO Contact from the Pleiades – A supplementary investigation report. – Annex IV* – page 546).

Dilettoso aside, the Wedding Cake UFO (Figure 2), on initial perception can all too easily suggest the Meier case is unreal or faked. At first glance, this UFO can look like a little model made with too many decorative and complicated bits and pieces. It, in no way, resembles the standard aerodynamic disk-like UFO of the beamship in Figure 1.

Using Christmas tree balls, a food container lid, and other household items, sceptics have made similar models to this UFO, which at first sight, might suggest Meier had done the same. If this is a real UFO or flying machine made by extraterrestrials, why is it so complicated and non-aerodynamic? This book and the 11 experiments in the book *Researching a Real UFO* (Zahi and Lock) show that irrespective of its shape and form, the WCUFO is a big object, the size of a compact car and that no matter it may look somewhat like a toy it is not one. Other photos, discussed in detail in Part I of this book, demonstrate other WCUFOs are twice the size of the one shown in Figure 2.

Figure 2 - Wedding Cake UFO hovering above Meier's front yard. Photo #808 in Meier's album.

On the cold, foggy, and snowy evening of 18 March 1975, the Plejaren made what is now called *The Pendulum UFO* film demonstration for Meier. Meier mounted his 8 mm film camera on his tripod and for a few minutes recorded a UFO or "beamship" as the Plejaren call their flying disks, dancing around a tree close to a house about 380 metres from the camera (see Figure 3).

Viewing this film for the first time can give the impression of a toy hanging from a thin cord attached to a fishing rod that Meier might be holding, that he moves back and forth and in circles. The

Introduction: The Worldwide UFO Controversy

tree, however, looks real and very close to the house, but our natural sceptical nature questions whether Meier has used one of two false perspective tricks. The first: a little UFO model suspended close to the camera, creating the illusion of appearing close to a more distant large house. Second: a small tree or bonsai (suggested by those who know nothing about bonsai) together with a toy hung from a thin cord very close to the camera with the house far away.

It is indeed a very curious film. Watching it, the UFO appears to move much as though it is hanging from a cord. However, the tree seems enormous, close to the house, and absolutely nothing like a bonsai. Furthermore, sometimes the UFO goes in front of and behind the tree, indicating it and the tree are very close to each other, at least at some point; and significantly no sceptic or investigator has ever precisely duplicated this craft's movements. A little UFO model cannot make the treetop move without touching the tree. Even a small tree is too big for a model to move without touching it. The tree in Meier's film only moves as a result of a right turn the beamship performs that causes the tree-top to move without touching it. It moves as if a significant object is causing turbulence, or the beamship emits some mysterious force field that causes the treetop to move. This specific movement remains unreproduced.

Figure 3 -   Pendulum UFO. A beamship dancing around a tree.

Figure 3 was taken from the "FULL Billy Meier-1985 Beamship - The Movie Footage" by Junichi Yaoi (1985). Yaoi went on to become a legend in Japan, producing an immensely popular series of hour-long UFO Special TV programs broadcast nationwide during the 1980s and 1990s. The seminal Nippon Television 1970s Meier interview in Japanese with Japanese subtitles made by the young Junichi Yaoi (now retired) covers professional top scientific photographic analysis performed at the time. It proves Meier did not use double exposures, and that the 7-metre craft reflects the distant landscape beneath it. It further reveals the inexplicable nature of the metal samples received and analysed plus a great deal more. Here is the link to this rare Japanese TV program: https://www.youtube.com/watch?v=xvbtGJKljQg. In Part II of this book, this film is discussed and analysed in some detail.

An extraordinary event happened to Wendelle Stevens with a copy of *The Pendulum UFO* movie footage that he obtained on one of his visits to Switzerland. Stevens covers the incident in his book (*UFO Contact ...: A supplementary ...* page 536). At his home in Tucson, Arizona, USA, in the fall of 1981 Stevens was showing the film to reporters from Denver. He did not have the authorisation to give copies of this movie footage to anyone, but he thought that allowing the reporters to film a few seconds would not present a problem. He projected the movie onto a wall, showing the UFO moving the tree-top while reporters were trying to record the image from the projection.

Inexplicably, after several attempts, the journalists' cameras and tape recorders consistently failed to work. Their equipment began behaving bizarrely. They could not record the film no matter how many times they tried. At the same moment in time as the showing, Stevens received a call from somebody saying that a UFO or visible moving light was flying above Tucson. Was this just coincidence, or was somebody up there in the skies over Tucson interfering with the journalists' equipment? If so, perhaps they concluded that many people were not ready at that time to see this evidence, whereas now humanity is.

Alternatively, perhaps the Plejaren were adding inexplicable happenings to the UFO controversy: creating more questions rather than providing answers; questioning appearing to be the Plejaren preference, questions in the form of, "What is going on?" that challenge us to find the answer.

Introduction: The Worldwide UFO Controversy

The Plejaren explained to Billy Meier that they moved this beamship in the film to imitate a pendulum, in order not to shock the psyche of the many people still stuck rigidly in their current, often religious, belief systems of the time. Maybe humanity was not ready then, almost 45 years ago, to accept the proof of a real ship from outer space. Ptaah, whom Meier says is an extraterrestrial Plejaren commander and human being with whom he maintains friendly and informative face-to-face exchanges, explains in Contact Report 251:

> For this reason, we executed maneuvers with our genuinely existing flying objects, e.g., jerky pendulum movements and certain skipping and floating movements, which ordinarily are not a part of our flying objects' normal movements and flight techniques. Instead, they appeared like those ridiculous, abrupt pendulum and strange swaying movements of the manipulated movies and photos by the UFO swindlers, liars and charlatans. The fact that we chose to execute these movement types is based on the reasoning that they, in particular, would result in the worldwide, fierce UFO controversy which resulted in many hardships for you, of course (Meier).

The UFO controversy was required, but it caused many problems for Billy Meier and adversely affected his reputation. So why did the Plejaren do it? Is humanity unprepared for open contact?

It is our considered opinion that these ETs know that the collective body of Earth humanity, are neither ready to accept the fact that we are not alone in the Universe nor that the ETs are, and have been here for a long time. Our think tanks echo the Plejaren view that humanity at large is not ready for extraterrestrial interaction.

The Brookings Institute, together with NASA, are likely withholding any knowledge of things evincing extraterrestrial intelligence while they continue assessing the likely impact of such a revelation. Brookings recommended to NASA in 1960 and presumably still recommends:

> ...continuing studies "to determine emotional and intellectual understanding and attitudes" regarding the possibility of

intelligent extraterrestrial life and studies to understand "the behavior of peoples and their leaders when confronted with dramatic and unfamiliar events or social pressures." The latter aimed to determine how such information might be shared, or withheld, from the public (Brookings Institute 2019).

Brookings' comments imply the withholding of information until they see it as safe to disseminate. Their "Proposed Studies on the Implications of Peaceful Space Activities for Human Affairs" is commonly known as "The Brookings Report." The report's penultimate section "The implications of a discovery of extraterrestrial life" mentions how such a revelation might adversely impact religions, the religiously minded, nations and states, as well as present threats to the roles of leaders and leadership groups (Brookings themselves?):

> Anthropological files contain many examples of societies, sure of their place in the universe, which have disintegrated when they have had to associate with previously unfamiliar societies espousing different ideas and different life ways; others that survived such an experience usually did so by paying the price of changes in values and attitudes and behavior (Michael page 183).

Furthermore, our think tanks obviously would not relish the idea of relegation to second place on the intelligence/leadership scale, given the presence of superior extraterrestrial intelligence. Their stated reasons for the prohibition, however, echo those of the Plejaren and are – or were – real enough. Namely, the dissolution of the lesser culture as the higher culture absorbs and rules over it; and psychological imbalance of the individual and collective psyches. Notably, however, the leaders of the lesser culture are the ones who lose most. The populace gains knowledge and benefits while the leaders lose many aspects of revered status.

In our investigations, we too, have come across people who have been severely affected by having to confront a real UFO experience, and the ET Plejaren know that an experience like this may adversely affect many people psychologically, leaving them worrying how to cope with the reality. While the Plejaren say they do not want to cause disturbance to the psyche of many people on

Introduction: The Worldwide UFO Controversy

Earth, at the same time, they know that we should, and will, one day make contact. We people of Earth are in a slow process of change and acceptance, and having a worldwide UFO controversy is part of this process. It allows people not yet ready to accept the reality of the existence of extraterrestrial beings an easy way out, by confronting a sceptic barrier that provides them with an exit door. Others, however, who are willing and ready to investigate the evidence in detail, can find their UFOs are real and know *They are Here.*

Another example of evidence that at first sight seems a hoax, but that after a detailed investigation we concluded was a composite of real and manipulated photos, concerns the Energy Ships. We may examine and publish this evidence in a future publication. According to Meier, the Energy Ships are flying spacecraft made of energy, not metallic or physical material. The Plejaren explained to him that the construction materials of spaceships from different cultures lead to their cataloguing them into three types: the metallic; the biological; and those of pure energy. The Plejaren use metallic beamships that can move into other dimensions and travel through time. However, in 1979 Meier says he received a visit and manifestation of these Energy Ships, another non-standard form of UFO.[1]

Figure 4 - Photo #728 from the Meier collection. An Energy Ship is hovering above Meier's parking lot?

They are Here

Sceptics have claimed these photos to be double exposure photographic tricks of taking a night shot of the car park, and another shot on the same negative of a bright object like a lamp, then overlapping the images to create the desired effect. They also found the edges of these ships are identical to street lights on Meier´s property, and so sceptics suggest, not surprisingly, that Meier used them to create the trick.

We found in some photos that these ships projected beams of light interacting with the surrounding environment by illuminating the landscape and cables below. Because these "objects" sometimes interact with the environment, we conclude some are mysterious flying objects, not a result of double exposure trickery, despite there being a triple exposure photo that the authors seem to have been among the first to notice as a triple exposure among these photos.

Our current conclusion on this multiple exposure photo (#720) is that it was most likely not made with the camera, but instead, was the result of darkroom work to produce just a colour print multiple exposure, that Meier subsequently received. By whom remains an intriguing question, as goes for other manipulated Meier colour photos, a very likely reason for their production, of course, being attempts by professionals or threats to darkroom or lab staff by people unknown who were determined to discredit the case.

Other "evidence" supplied by sceptics as "proof" the Meier case is not real are some fake photos from Meier's trip known as his Great Journey. Meier says he made a great journey across the universe with the Plejaren in which he shot several rolls of photos far out in the universe beyond our solar system, even in other galaxies. Unfortunately, forgeries replaced these original pictures, so, Meier received the rolls developed with cleverly disguised fake paper copies of doctored images that looked remarkably similar to the incredible photos he had taken. Somebody, possibly a "Mr Schmid" among others who helped send Meier's film rolls to a lab, for diverse reasons, including alleged threats to his livelihood or life, allowed manipulation of the photos by people interested in damaging Meier's reputation. Meier now says he did not receive the original photographs from these film rolls of his trip.

Initially, neither Meier nor the Plejaren noticed the subtle changes in the imagery. After their detection, however, the Plejaren

Introduction: The Worldwide UFO Controversy

destroyed most of the photos saying they would only bring Meier more trouble. Meier kept a few of these pictures, but it is not clear what level of manipulation was performed on the remaining ones. It is common, therefore, to find these photos faked by unknown persons from books and movies, not from outer space, going around the market place. There are around 40 such photos out of the 1,378 Meier has or had in his collection. Examples of the fake photos that he knows are not originals are the "Asket & Nera" pictures, the dinosaur pictures, the Apollo-Soyuz photos, the images of the universe barrier or gateway to a neighbouring universe, some deep-space photographs and more.

One way or another, the evidence presented gives room for doubt, and it certainly offers an immense trove of time-consuming material to investigate for those new to studying this case. It might, then, look faked based on initial or surface perception rather than a rational and thorough investigation. Moreover, the Plejaren themselves actively contributed to this deniability. Deardorff's enlightening paper "Plausible Deniability" discusses several logical hypotheses that cover possible reasons and causes for this widespread phenomenon of never having conclusive evidence for UFO cases (2013).

Every day the Meier case evidence is reviewed, more hidden clues are found within it. Within its abundant information and extraordinary evidence of encyclopedic depth, humanity can finally realise, hopefully not too late and before the Plejaren leave for several hundred years, that we are not alone, and that we are fully responsible for our destiny. We need to evolve our consciousness to a level that enables contact with advanced space-travelling cultures on a face-to-face basis, without us feeling threatened by them or being a threat to them. This process has probably happened on other planets similar to Earth, and with their broad spectrum of planetary experiences and interactions throughout many millennia the Plejaren doubtless would know well, both how, and who is best among us, to chaperone such a change and process.

As one small step in this process of change and acceptance, this book presents an invitation to check some of the compelling evidence of this extraordinary case and decide for yourself whether it is real or not. It is time to confront this issue and seminal case and determine whether or not it is part of our reality. The authors

They are Here

of this book have decided it is. Your decision on this, the most significant UFO controversy of them all, is up to you. Whatever your conclusion, enjoy your read.

Part I

The Wedding Cake UFO - WCUFO

Introduction to the WCUFO

Billy Meier, a Swiss country dweller, claims he has been in contact with extraterrestrial beings called the Plejaren who come from a place in a different space-time configuration 80 light-years beyond the Pleiades open star cluster located in the Taurus constellation. Meier says they are human beings, like us, but with a higher level of technological and spiritual evolution. Meier claims he has taken over a thousand photos of their craft and things he witnessed on journeys with them. Hundreds of the pictures survive. Some show a strange UFO sporting three or four dozen reflecting spheres and various minor parts, like coloured lenses and coloured crystals. Meier took a few dozen photos of this particular UFO, which because of its peculiar form, has been called the "Wedding Cake UFO," or WCUFO in print. Meier refers to it as the "cake UFO."

Why is this WCUFO so complicated, and why does it not have the aerodynamics of a typical flying saucer? Maybe a hint to answering this question lies in one of Meier's photos, of another UFO, called the "Plejaren miniscout" (Figure 5, bottom UFO), that sports three undercarriage spheres.

Figure 5 - Billy Meier's photo, showing a Plejaren beamship above and a miniscout below sporting three undercarriage spheres.

They are Here

Do most of the Plejaren beamships, have some form of these spheres? The miniscout shows three of them visible on its undercarriage, but the WCUFO displays around 43 spheres. In other beamships, like the one above the miniscout in Figure 5, these spheres are not in evidence, perhaps because they withdrew inside the craft.

Because the WCUFO was only for use in Earth's atmosphere, could its spheres constitute a small sophisticated engine with an internal mechanism producing a magnetic field that interacts with Earth's magnetic field? If so, it seems to us that by having a few dozen spheres, the magnetic field it produces may give it stability in our environment. A noticeable difference between the WCUFO and other beamships Meier recorded on video is this WCUFO's stability in the earth's atmosphere. In essence, it hovers completely stationary, while the other beamships present a wobbling movement, as though they are floating on the earth's magnetic field. Whatever, the WCUFO creators may well be showing us the inner workings of one of their ships. If so, this WCUFO is a UFO just like the other beamships minus an outer cover.

Consider ourselves as future space travellers arriving on another inhabited planet. Wishing to promote ideas, rather than present specific knowledge among the inhabitants, we build and reveal a vehicle capable of working in their environment, which could inspire them to progress and construct an automobile. We remove, however, the auto cover to expose the core engine and mechanical parts to their budding scientists. Perhaps, similarly, the WCUFO creators wanted us to learn through personal efforts to discover the functions of the various WCUFO parts and how to construct a similar ship.

The WCUFO may provide a technological missing link between our current vehicles and space travelling UFOs: a necessary technical step between our earthbound vehicles and vehicles that traverse the universe.

Initial designs of this WCUFO, allegedly coming from the 1920s, suggest this possibility. Contact Report #254, which Meier says he transcribed from his conversations with the ETs, also supports this narrative. Ptaah further explained the origin of the initial WCUFO designs to Meier in this extract from the Contact

Report including the original German text (28 November 1995; FoM 2015):

Ptaah:

4. We already worked with those flying devices, which you call the cake-ship, in the twenties, but it was indeed only at the end of the seventies that they were brought to the required status for their use on the Earth.

5. The form of these flying devices was specially thought up for the Earth, for which reason we made the effort to transmit the entire necessary specifications for the design to terrestrial scientists through impulse-telepathy so that, out of that, flying disks could be developed.

6. This impulse-telepathic information went predominantly to aerospace technicians, as I will designate these persons, whereby especially German engineers were included for this, to whom we transmitted exact plans for the external form as well as certain technical particulars which were responsible to transmit.

7. Thereby the German scientists also actually experimented, whereby they could construct halfway suitable flying disks, which according to our thinking at those times should have been used to constitute an air power through which an early-brought-about world peace should have been achieved.

8. However, the political machinations changed very quickly into a bellicose direction, for which reason we brought an end to further impulse-telepathic information to the German

Ptaah:

4. Bereits in den zwanziger Jahren arbeiteten wir mit jenen Fluggeräten, die du als Tortenschiff bezeichnest, doch für den Einsatz auf der Erde wurden sie erst Ende der siebziger Jahre auf den erforderlichen Stand gebracht.

5. Die Form dieser Fluggeräte war speziell für die Erde gedacht, weshalb wir uns auch bemühten, impulstelepathisch die gesamten notwendigen Angaben für die Form an irdische Wissenschaftler zu übermitteln, damit daraus Flugscheiben entwickelt werden konnten.

6. Diese impulstelepathischen Informationen gingen vorwiegend an Weltraumfahrttechniker, wie ich diese Personen bezeichnen will, wobei besonders deutsche Ingenieure dafür einbezogen wurden, denen wir genaue Aussenformpläne sowie gewisse technische Einzelheiten, die verantwortbar waren, übermittelten.

7. Damit experimentierten dann die deutschen Wissenschaftler auch tatsächlich, wodurch sie halbwegs taugliche Flugscheiben konstruieren konnten, die unserem Sinn gemäss damals dazu benutzt werden sollten, eine Luftkraft zu bilden, durch die ein frühzeitig herbeigeführter Weltfrieden erlangt werden sollte.

8. Die politischen Machenschaften jedoch veränderten sich sehr schnell in kriegerische Richtung, weshalb wir von weiteren impulstelepathischen Informationen an die deutschen

scientists and allowed the project to expire, whereby we however initially transmitted false information so that the flying disks could not be created specifically for warlike purposes.	Wissenschaftler absahen und das Projekt fallenliessen, wobei wir jedoch erstlich noch Falschinformationen übermittelten, damit die Flugscheiben nicht zweckgerichtet für kriegerische Zwecke erschaffen werden konnten.

According to this report, the original designs of these flying machines were for craft intended to operate on Earth. Unfortunately, due to the aggressive scientists and politicians, Earth lost the opportunity to develop flying disks like these. To date, our scientists have not been able to construct them, at least in public, but despite this setback remarkable photographs of them exist.

Figure 6 - Photo #808 shows a 3.5-metre diameter WCUFO with crystals, lenses, golden features and various little parts. Meier's home is behind.

On 22 October 1980, Meier saw and photographed a WCUFO hovering above his courtyard. Previously, he had been taking

Introduction to the WCUFO

photos and videos of several flying disk type beamships. Now, he says the Plejaren first visited him with a WCUFO: a 3.5-metre diameter flying ship. In Figure 6, a fantastic photo of this WCUFO shows numerous coloured crystals and small lenses. A blue one among them is inside or capping one of the spheres.

The house in the background on this cold day in Figure 7 is Meier's home. In the sphere reflections, a building is seen behind Meier when he took this photo. This building is the wooden carriage house enclosing the parking lot or courtyard opposite Meier's home. Momentarily we show that Meier was very close to the carriage house wall while this small 3.5-metre diameter WCUFO was flying in front of and quite close to him at a bit less than six metres from his camera.

Figure 7 is a wide-angle view of Meier's main property showing his house on the right and the carriage house on the left. The WCUFO flew close to where the orange car is in the middle of the parking lot. There were, however, no automobiles in the parking lot on the day Meier took his WCUFO pictures.

Figure 7 -   Wide-angle view of Meier's main property in 1981.

Figure 8 shows the eleven photos Meier took this day, the picture numbers coming from his photo albums. Part I of this book presents a summary of studies conducted on some of these photos.

Later, on numerous occasions throughout almost one year other WCUFOs are said to have come to Meier, and he was able to take additional pictures and a video of one of them.

They are Here

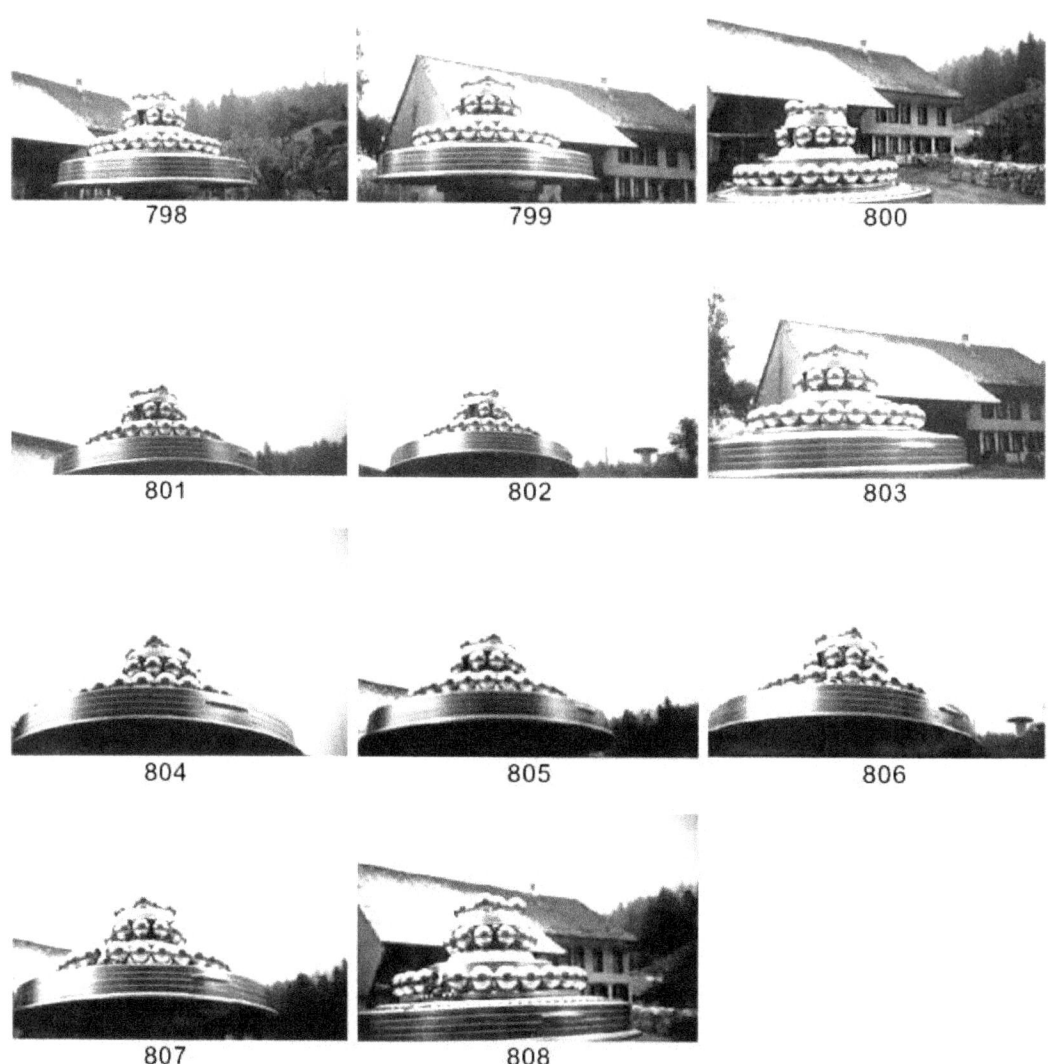

Figure 8 - Eleven photographs of the WCUFO taken by Billy Meier in his front courtyard on 22 October 1980, at 11:23 am.

Introduction to the WCUFO

Figure 9 -   Photo #829, 26 March 1981, at 6:19 am. A 3.5 or 7-metre diameter WCUFO.

Figure 10 -   Photo #834, 3 April 1981, at 1:10 pm.

On 26 March 1981, Meier took several photos of a 3.5-metre WCUFO between him and his green trailer. Only two came out successfully (Figure 9). The front of the WCUFO is just out of focus, and its middle, the trailer, and the horizon are clearly in focus. This book presents a detailed updated analysis of this photo, significantly expanding on the short report in *Researching a Real UFO* (Zahi and Lock page 7).

They are Here

Figure 11 - Photo series among the treetops.

A week later on 3 April 1981, Meier says he was flying among the treetops at around 1:10 pm on a 7-metre WCUFO taking photos of a companion 3.5-metre WCUFO. He shot several pictures of this WCUFO behind tree branches, Figures 10 and 11.

Meier says Quetzal, another Plejaren commanding these ships, accidentally broke off the upper portion of one tree, a photo of which exists (see Figure 53).

Analyses of this WCUFO sphere's reflections prove it is a sizeable object, not a small scale model. A scale model UFO would have to be proximal to the nearby tree on the left, and the tree reflection in the spheres would then cover a significant area of the spheres, but it does not.

A few minutes later the 3.5-metre WCUFO stayed motionless immediately in front of a 6 to 8-metre tall Norway Spruce tree. Meier took a series of photos as he walked towards it on the ground. Figure 12 shows a shot from the series. "Brightness" enhanced it reveals a vast shadow the craft casts over the tree's sizeable branches. Notice also the large pine needles, and the dark reflection on the WCUFO's left side undercarriage confirming its proximity to the towering spruce. Our calculations based on the camera Meier used, put his location at around 16 metres from the craft.

Introduction to the WCUFO

Figure 12 -   Photo #844, 3 April 1981, at 1:35 pm. A WCUFO casting a massive shadow over a young Norway Spruce.

One hour later, at 2:33 pm the same day, Meier took another series of photos and a video of this WCUFO close to another tree. Figure 13 (top) shows a shot from the series. Image processing it (bottom) reveals the branch and leaf structure of the large tree in front of the WCUFO.

On the night of 2 August 1981, Meier says the WCUFO returned. The pure silver colour now looks golden at night. In this book, an analysis of its bright sphere "reflections" suggests that rather than reflecting street lights it *emits* light. Figure 14 shows this 7-metre WCUFO behind a Mercedes-Benz vehicle, and a little tree closer to the camera. The tree is totally out of focus and the car a bit blurred due to the camera's short depth of field in this night photo. Meier took several WCUFO photographs in this location very close to the parked Mercedes-Benz car.

They are Here

The trace of light on the top left is said to have been caused by a small telemetric disk scurrying around during the photo exposure. Telemetric discs, according to Meier, are little flying monitoring devices the Plejaren use.

Figure 13 - Top: Photo #850, 3 April 1981, at 2:33 pm. Bottom: The same photo image processed to enhance and reveal this Norway Spruce tree in front of the WCUFO with its bright red crystals.

A few days later on 5 August, again at night, Meier took what many now think is the most remarkable of all WCUFO pictures: Photo #873 (Figure 15) showing a WCUFO with its central core

Introduction to the WCUFO

extended upwards by about 13 centimetres. A few years ago, we thought he took this picture while flying on another WCUFO in a central area of the Swiss countryside. Christian Frehner, however, informed us by email (March 2020) that Meier took this photo at the SSSC behind his home. Enhancing the brightness of the picture renders visible the previously hidden terrain and a fence pole below the WCUFO.

Further enhancing of this WCUFO image reveals a strange violet or red wine halo around it that the findings of an ex NASA scientist indicate is a classic atmospheric plasma sheath. Remarkably these details had remained undiscovered for over 30 years. Further detailed picture analysis suggests this WCUFO can hide the light it emits by employing a screening field around itself. More WCUFO photographs are available in the book *Photo-Inventarium* (Meier).

Figure 14 -   Photo #999, 2 August 1981, at 2:18 am.

They are Here

Figure 15 - Photo #873, 5 August 1981, at 2:48 am. A classic plasma sheath.

We can find similarities between the parts of this WCUFO and everyday household items. The WCUFO base looks like an available rubber or plastic trashcan lid, and additional components are similar to other household items. It seems that after Meier photographed this WCUFO, for reasons unclear, some manufacturers made different everyday items almost identical to many WCUFO components in a scale befitting a 55-centimetre diameter model. Not surprisingly, therefore, would-be debunkers support the unproven contention that the WCUFO is a small model and that the WCUFO case is a hoax. Hence the WCUFO has successfully helped embed the planned Worldwide UFO Controversy explained in this book's Introduction. However, our investigations and the indications in this book and our book *Researching a Real UFO* prove that this and other WCUFOs are indeed large objects, of at least two sizes: approximately 3.5 metres and 7 metres in diameter.

Billy Meier's Courtyard WCUFO: A Detailed Analysis

Introduction and key findings

On 22 October 1980, Billy Meier began taking pictures of the first WCUFO using a Ricoh Singlex TLS camera, with a focal length of 55 mm. Both Christian Frehner and Billy Meier said until February 2020 that they think he used the Olympus camera for these photos. Still, we show why he must have test used the Ricoh for just these first WCUFO photos, and then presumably set it aside due to his difficulty in using it with only one arm, especially in the surrounding mountainous countryside.

This UFO was not like the others he watched and photographed before. The new UFO had a different look, with many peculiar details, including many spheres encircling it. Because of its appearance, it was dubbed "the wedding cake UFO" (WCUFO). Meier usually refers to it as the Cake UFO.

Here, some WCUFO photos taken by Meier in the parking area or courtyard of his property are analysed. Later in this book, other WCUFOs hovering above the treetops, and some nighttime WCUFO photos are analysed.

For his 22 October 1980 courtyard photos, Meier used a thick piece of glass in front of his camera to prevent adverse effects due to its proximity to this UFO. This glass may have caused the resulting sphere reflections to be a bit blurred; however, our investigation finds that despite any effect of the glass, reflection analysis allows a very reasonable estimate of the WCUFO's size.

We conducted several analyses using two methods: Steel reflecting spheres in a scale model of Meier's property; and a computer analysis using the widely used modelling tool Blender to create 3D models and animations. Summaries do not delve into too much geometry and math to make this present book easy to follow. [2] Recently other investigators, like Steve (Lane) and "Taro" (Istok), have found similar results from excellent modelling of the WCUFO and Meier's site.

They are Here

To better understand the location's surroundings and background, images of the dark shapes reflected on the WCUFO spheres were processed. In our book *Researching a Real UFO: A Practical Guide to WCUFO Experimentation for Young Scientists*, 11 experiments were introduced that anyone can perform as part of a practical science project to find many WCUFO details. To assist those requiring more WCUFO details, these are the 11 experiments:

1. Understanding spherical reflections
2. Making a plan view of Billy Meier's property and site (the WCUFO photographed location)
3. Establishing Meier's location
4. Making and using a scale model of Meier's residential area
5. Making and utilising large scale models of Meier's site & camera
6. Calculating the WCUFO angle of view and size
7. Distance estimation using the camera formula
8. Mapping the WCUFO's local environs
9. Image processing photo images to discover hidden details
10. Making a stereoscope for viewing 3D WCUFO pictures
11. Height changing capability of a WCUFO

Our models and computer analyses lead us to conclude:
- The WCUFO photographed on 22 October 1980, hovering above the main parking area of Meier's property, has a diameter of between 3.0 m and 3.6 m. (The Plejaren reported it to Meier as 3.52 m).
- This WCUFO cannot be a small-scale model as some sceptics claim. It is not a model made from a 55 cm diameter trash-can lid, or even a bigger one-metre diameter model. Several tests and modelled possibilities are conducted locating this WCUFO in different places within the courtyard area. These tests and models lead

us to conclude that it is impossible to take such pictures using a small model. This WCUFO is an object larger than three metres in diameter and probably is a 3.5-metre object just as Meier says he was informed. [3]

- The WCUFO has different proportions depending on its size. Specific photographs of apparently different WCUFO sizes are analysed. We conclude the horizontal proportions of the 3.5-metre and 7-metre diameter WCUFOs are the same. The vertical proportions, however, can differ somewhat and this, of course, discredits any single small model claim. In Meier's home parking area, only the 3.5-metre WCUFO appeared.

- Stereoscopically viewing the reflected dark carriage-house shapes in 3D provides further information. Forms of various objects appear at different distances. Among 11 photos taken in front of Meier's house, photo #808 is the clearest because the WCUFO was then at its closest to the camera. Significantly for analysis, it also displays probably the best reflection of the carriage house wall.

- The many little details found in picture #808, like its different coloured crystals, coloured lenses, and engraving details in its several parts make this WCUFO extraordinarily challenging to reconstruct precisely as a small model (see Figure 6, photo #808 detail). Constructing these on a model of just 55 cm would require intricate miniature work.

Our investigation analysed only a few of the many WCUFO photos. The others, plus a video, remain available for further research by interested parties.

They are Here

Photo #800 analysis

Meier took eleven photographs of the WCUFO hovering just above the main parking area or courtyard of his property at 11:23 am on 22 October 1980 (Figure 8). One of them, #800, was analysed to determine the WCUFO's actual size. Some would-be debunkers claim Meier made a scale model using household items, including a 0.55 m bin lid which they say he photographed. This claim was found incorrect. Checking the dark shapes reflected in the WCUFO spheres conclusively shows that this WCUFO is a much larger object.

The foreground of the photo is in clearer focus than the blurred background, which tells us Meier did not use his Olympus 35 ECR with its focus jammed on infinity for this shot.

Figure 16 -   Photo #800, WCUFO with Meier's house in the background.

Figure 16 shows the WCUFO photograph #800. Meier's house is visible in the background. Every sphere shows a very similar central dark shape. Figure 18 shows a zoom image of two central spheres on the lower tier of this WCUFO. These two adjacent images observed through a stereoscope show in three dimensions that the dark shapes appear to be distant trees on either side of a building. The central dark shape corresponds to the northeast wall of the carriage house. Meier was standing very close to this wall when he took the picture. See the book *Researching a Real UFO* (Zahi and Lock) with instructions on how to construct a stereoscope, how to make a plan view of Meier's property, and other experimental details referred to earlier, and here.

To determine the size of this WCUFO required knowing:

(1) The plan view or map of Meier's site. We made measurements on-site and used Google Earth images. (See Figure 17.)

(2) Meier's location. His location was found using three methods which proved he was up close to the carriage house wall to within an error of 50 cm in the photos analysed:

(a) the orientation of the background house in the photos

(b) Meier's distance from a specific home detail calculated using the *camera formula*.

(c) whether the photos show what they must show given the type of lens and camera Meier was using

(3) The distance from the camera to the WCUFO. This distance was calculated for a central reflecting sphere on the lower tier of spheres, which gave a good indication of the distance to the WCUFO. The size of the carriage house wall reflection on the sphere images gives this distance. It is shown that for this photo, the central sphere was situated around 6 ~ 6.50 m from the camera.

Figure 17, a plan view of Meier's site, shows with a little dot Meier's position close to the carriage house wall SW of the

They are Here

WCUFO. The circle illustrates the WCUFO location and the construction in the upper right is Meier's house shown in the background of photo #800. Points A and B are respectively, the WCUFO central axis and its right edge projection toward the house.

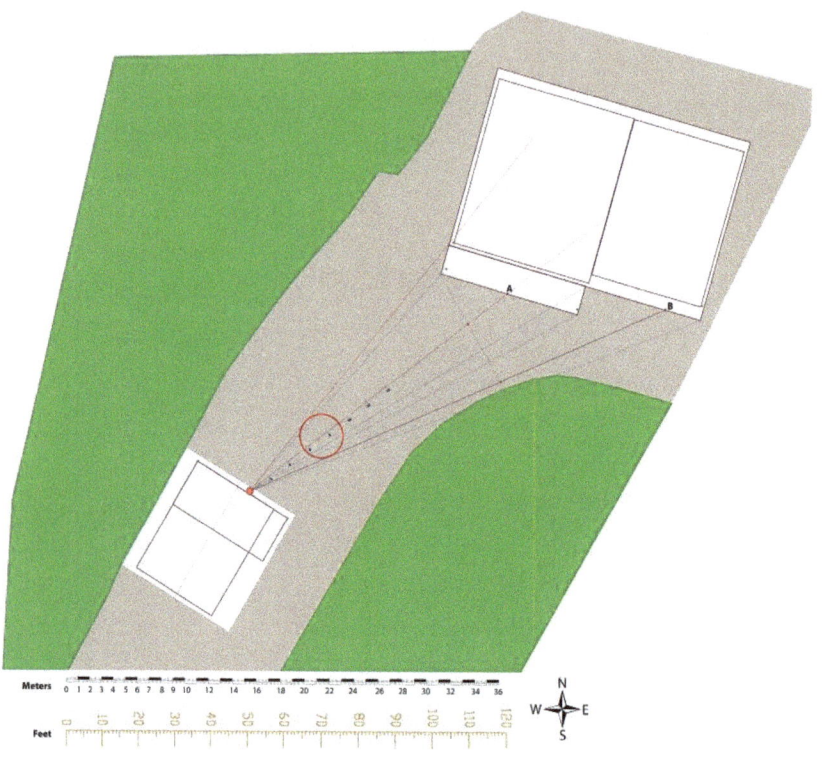

Figure 17 - Plan view of Meier's residence and courtyard in 1980. The circle indicates the WCUFO location in photo #800 with its central axis A.

This plan view now enables analysis by using the following two methods:

1. **Sphere reflections in a courtyard scale model**: Creating a scale model of 1 metre to 5 centimetres of Meier's property, including the carriage house wall, and taking pictures of a small reflecting test sphere at several distances. Checking the reflected image of the carriage house in the test sphere photos to determine its distance from the camera. Then

checking the size of the model WCUFO-sphere, and hence the scaled-up actual WCUFO size. Full details on how to make this scale model of Meier's property, and also how to create and use a large scale model to confirm the WCUFO size are found in our book *Researching a Real UFO*.

2. **Computer Modelling**: Creating a computer 3D model with the widely-used free 3D modelling and animation software tool "Blender" and assigning a reflecting material to the 3D WCUFO model spheres. Also, making a simple Blender model of the carriage house and Meier's primary residence. The WCUFO model need not be a detailed representation. The WCUFO bottom parts being unnecessary for the analysis, are not modelled. What is necessary is ensuring the precise positioning of all the spheres and their reflections.

Figure 18 shows zoomed images of the reflected dark shapes in two front bottom-tier WCUFO spheres. While initially appearing as mere dark shapes, viewed through a stereoscope, or naturally as some people can do with the naked eye, distant trees and the carriage house wall with nearby objects appear. These objects help distinguish the wall's approximate outline. Since these dark shapes are blurred, they extend laterally into a wider area, but the 3D view helps make possible a more precise delineation of the wall.

It is easier to identify the width of the carriage house roof in these blurred images rather than the wall itself. So in our analysis, the roof extension with eaves was used, which horizontally, is 10.3 m wide. The roof's reflected image extends a given horizontal percentage of the sphere diameter, which we found by measuring the roof extension in this image against the sphere diameter to be 34.9%. So the reflected carriage house roof width is 34.9% of the sphere's width. The 34.9% ratio is critical because if the WCUFO were a little 0.5 m model made from a trash-can lid, it must be no more than a metre away from the camera, making the reflection of the carriage house much more extensive than what the photo shows. The reflected roof would cover much more of the spheres' reflected image; precisely 60% of the sphere's width as opposed to 34.9%.

They are Here

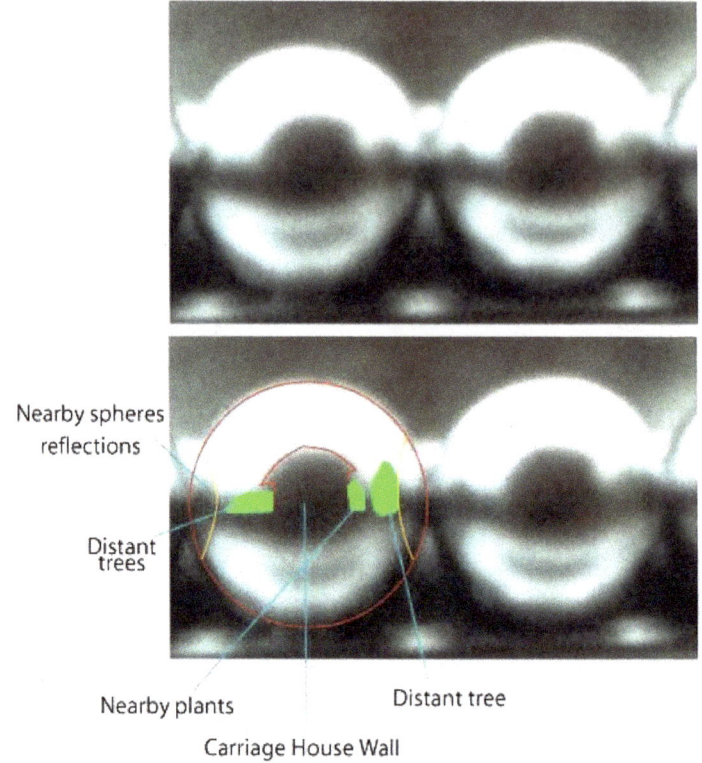

Figure 18 - Details of the reflected dark shapes on photo #800 front bottom-tier spheres.

Figure 19 left, shows how the carriage house reflection is if Meier used a little WCUFO model. The reflection of the wooden construction is now enormous. On the right, the sphere is six metres from the carriage house wall or the photographer's location. The reflected image on the right side is much closer to what Meier's photos show, revealing again that Meier did not use a little model. This figure comes from additional work and tests done on a larger scale model, as shown in Experiment 4 in *Researching a Real UFO* (Zahi and Lock pages 41 ~ 57).

Figure 19 - Sphere photos taken from the model of the carriage house wall. Left: Sphere photographed at one metre from the photographer reveals a much larger reflection than in Meier's photos. Right: Sphere photographed at around six metres away; the red rectangle inset shows a reflected carriage house image very similar in size to the reflected image in Meier's photos.

Method 1- Sphere reflections in a courtyard scale model:

With a plan view and detailed measurements of the carriage house, we conducted an experiment with a scale model of Meier's site, with the carriage-house northeast wall made of cardboard (Figure 20), and a reflecting high precision steel sphere at different distances from the camera. The reflecting sphere was located along the central axis A shown in plan view in Figure 17.

Figure 20 - Scale model of the carriage house northeast wall with a rectangular hole to locate an iPhone lens.

They are Here

Pictures were taken of the steel sphere atop a wooden rod. See the resulting photos at different distances in Figure 21.

The percentages in this figure are the width ratios of the carriage house roof to the sphere's diameter.

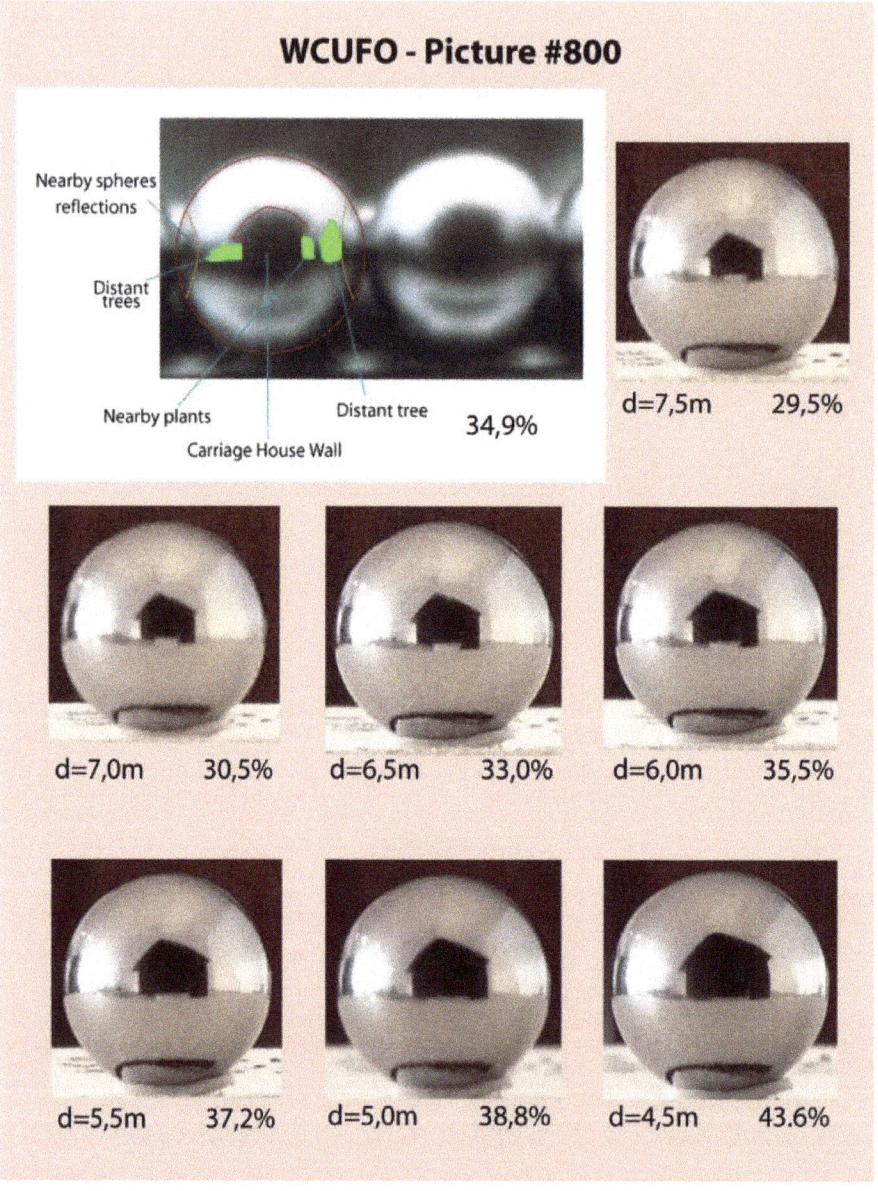

Figure 21 - Reflections of the carriage house model on a steel sphere (European numeral system of commas for decimal points).

At the bottom of each photo is the distance (d) of the test sphere from the carriage house wall, and the ratio of the roof width to the sphere diameter expressed as a percentage. Notice Meier's photo shows a percentage of 34.9% and the test sphere shows this percentage at between d=6 m, and d=6.5 m. The sphere was estimated to be at 6.12 m.

This distance now allows calculation of the WCUFO size, which is more than 3 m and probably nearly 3.5 m, as described in detail in the investigative *Analysis of the Wedding Cake UFO* (Zahi).

Some considerations:

- The 6.12 m distance found between the carriage house wall and the sphere has an error margin of no more than 5%. Whether Meier used a little UFO model or whether the object was a big WCUFO, the central sphere *must* be at this distance from the camera.

- As demonstrated in Experiment 1 in *Researching a Real UFO* (Zahi and Lock), the size of the sphere does not matter. In a test, either a small or large sphere is acceptable. The proportions of the reflected images are the same for any given distance. Wherever the photographer may be on the central axis A between the carriage house and the WCUFO in Figure 17, the size of the carriage house wall's reflected image occupies the same proportion of the sphere ("Sphere Reflection Rules" Rule c. page 158). [4]

  So a little model located at six metres from the carriage house produces the same reflection as the one seen in Meier's photos. Therefore, in this little-model scenario, Meier must be located at five metres from the wooden construction, but we know he was up close to its wall.

- Some sceptics suggest taking a little WCUFO model to Meier's premises and testing it. Of course, they are welcome and encouraged to do so, but it is unnecessary and redundant because providing the distance does not change, testing one single sphere of any diameter produces the same results as looking at the spheres of a little model. Moreover, it makes for better measurements and accuracy to use a more extensive sphere or model.

Earlier, Christian Frehner experimented using just such a big sphere on Meier's site. (Figures 22 and 23)

Figure 22 -   Reflecting sphere at 2 m from the camera, taken from a position close to the carriage house wall looking north. Right: Zoomed image.

Figure 23 -   Reflecting sphere at 1 m from the camera showing the photographer at the centre of the sphere. Right: Zoomed image.

In this experiment, a significant percentage of the sphere width is occupied by the reflected carriage house image, bigger than 34.9%. The reflected image is even more enormous if the Test Sphere is at only one or two metres from the camera; as shown earlier, it must then be around 60% of the sphere's width. Furthermore, if Meier used a little scale model of the WCUFO, it had to be just one metre from the camera.

Figure 24 - Confirming the WCUFO location during Meier's photo session; Erhard Lang holds a 3.5-metre long pole in the SSSC parking lot (the courtyard).

At Meier's courtyard (Figure 24) Erhard Lang performed another little exercise holding a 3.5-metre bar, the size of the WCUFO, at the correct distance away in the parking lot but shifted to the right of photo #800. The red arrow shows the correct position for the WCUFO in photo #800. If Meier used a little 55 cm diameter model, in this photo, its size would be the width of Lang's shoulders, not the size of the pole his is holding. It would have been minuscule in Meier's photos, not the big UFO we see there.

They are Here

Method 2 – Computer modelling

We constructed a computer model using the tool "Blender" which enables testing of different sizes and different WCUFO positions in Meier's courtyard. It also enables checking of the images reflected in the spheres.

Figure 25 shows our Blender perspective model of the WCUFO, the carriage house, and Meier's house. We omitted the unnecessary WCUFO base because only the sphere reflections are required. The configuration and sphere positioning's, however, are accurately portrayed.

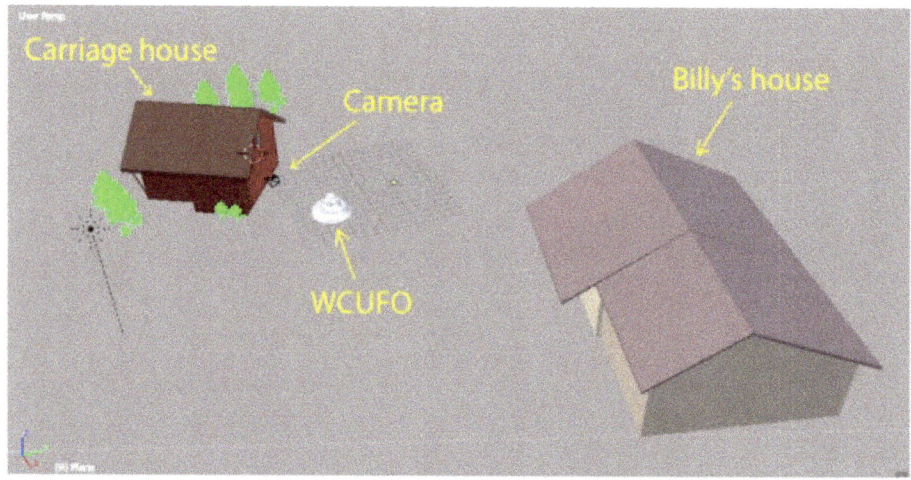

Figure 25 -   "Blender" computer-made perspective model of Billy Meier's courtyard.

The camera in this Blender model is close to the wall and indicated by a tiny black pyramid. Testing different sizes of the computer-generated WCUFO at different distances, the images "rendered" produced the reflected images shown in Figure 26.[5] The dark shapes in the spheres are very similar to those in photo #800, all except for those in the small 55 cm scale model.

The first clear and obvious conclusion is that this WCUFO was not a scale model like a 55 cm diameter model proposed by various sceptics who have never taken the trouble to test their claims. The carriage house wall in Figure 26 bottom right is grossly oversized with the camera at 1.16 m from the model. The roof extends 56.8% of the sphere's diameter, while the much smaller value of 34.9% is the correct percentage in photo #800.

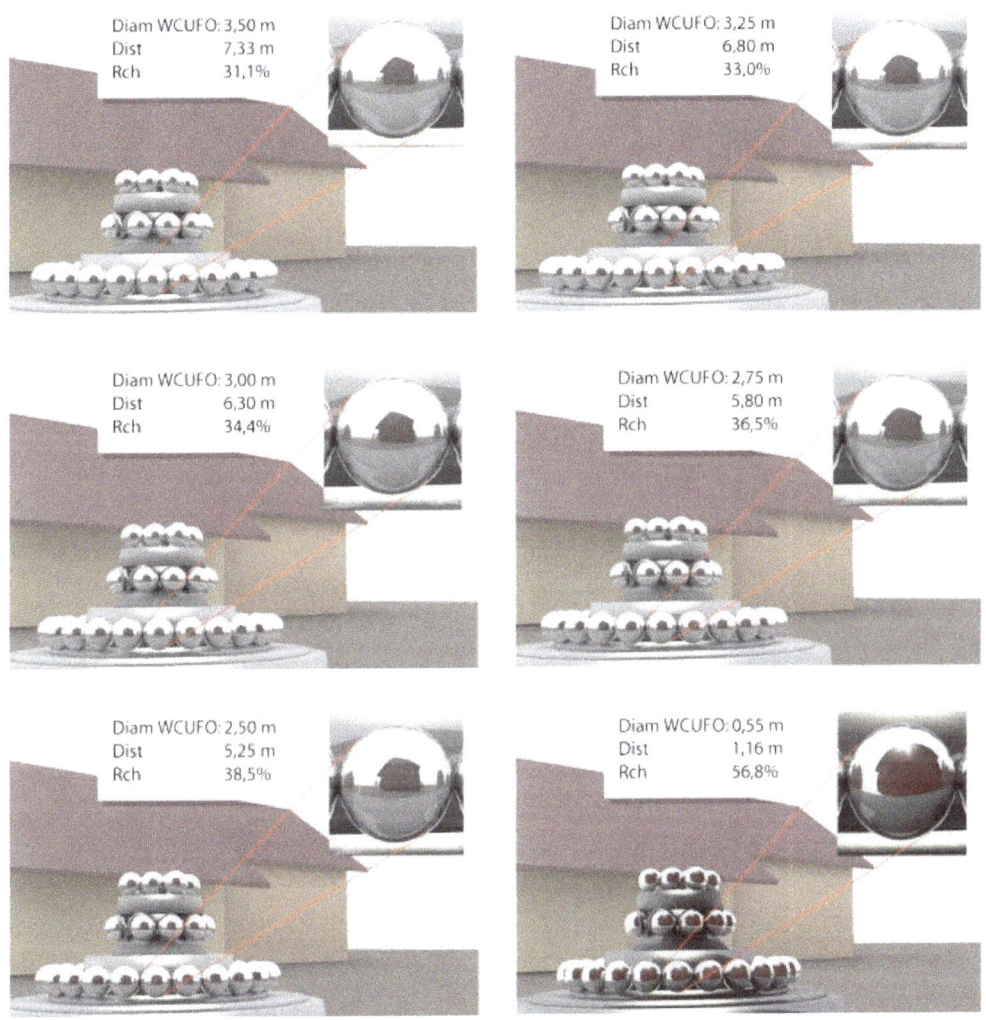

Figure 26 - Rendered WCUFO images at different distances and sizes.

If the roof width is 34.9% of the sphere's width in photo #800, then the 3 m diameter WCUFO image is a very close fit. So this method of calculation shows the WCUFO might be around three metres in diameter, or a bit less, not about 3.5 m as indicated in Method 1.

This tool and method enable the creation of a computer-rendered process. It remains unclear how precise this "render" method is for producing sphere reflections. Rendering in Blender may be quite suitable for modelling 3D objects and animating them, but the Blender render tool might not be very accurate in showing real sphere reflections. Method 1, using reflecting spheres in a scale model of Meier's property, probably presents the more precise results.

We tested additional photos like #799, found in Zahi's 74-page investigative report, *Analysis of the Wedding Cake UFO*. Our previous book *Researching a Real UFO* shows further tests and analyses performed on Picture #808 all of which show the same results: The WCUFO is an object bigger than 3 m in diameter, and perhaps around 3.5 m as Meier has always maintained.

Remarkable details in photo #808

Photo #808, Figure 27, shows some extraordinary features in clear focus, and the WCUFO positioned to display crystals of different colours. It also has intriguing engraving work, which, if this were a scale model of half a metre in diameter, would make for very complicated and precise, while perhaps not impossible, work.

Figure 27 - Billy Meier photo #808 in clear focus displaying remarkable details.

The background in photo #808 is the same as the other photos taken in Meier's courtyard. The WCUFO is now closer than in photo #800, covering 82% of the image width compared with around 70% in #800. The reflected image of the carriage house wall is a bit bigger and has a higher resolution. Again, as in photo #800, the foreground of the photo is in better focus than the blurred background, which tells us Meier did not use his Olympus 35 ECR with its focus jammed on infinity for this shot.

They are Here

Reflections on the spheres.

Figure 28 shows a zoomed image of spheres at the centre of this photo. On the right side, in yellow, the carriage house wall is delineated as it would look with spherical distortion. Once again, this WCUFO is around 3.0 to 3.5-metres in diameter, and not very close to the carriage house. In this case, Meier's size in the reflected image is tiny and indicated by the green silhouette. Only with difficulty and in excellent reproductions does the dark human figure stand out against the dark carriage house wall and its shadows.

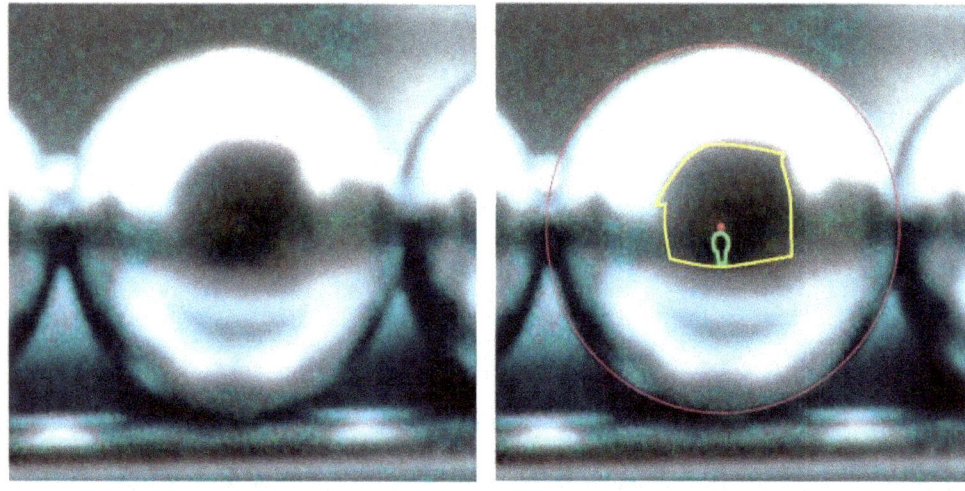

Figure 28 - Photo #808 central sphere details. Right: a polygon approximately delineates the carriage house wall. Delineated at the very centre is the inevitable tiny reflection of the photographer.

As indicated before, if this WCUFO were a scale model 55 cm in diameter at the required approximate one metre from the camera, and if Meier is close to the carriage house, the wall reflection will cover most of the sphere, like the reflection shown in Figure 23. These Figure 28 sphere images show Meier close to the carriage house wall and a fair distance away from the WCUFO.

The WCUFO sphere material seems not polished shiny, unlike reflecting Christmas tree balls, and the vertical base section is unpolished.

Billy Meier's Courtyard WCUFO: A Detailed Analysis

Coloured crystals and lenses

Remarkable details such as crystals and lenses of different colours reveal themselves in this photo. A zoomed image of this WCUFO's central region unveils vibrantly colourful red, blue, green, and two white crystals one on either side of the bright light blue crystal formed in a tear-shape extending down over the rim of the WCUFO base (Figures 29 and 30).

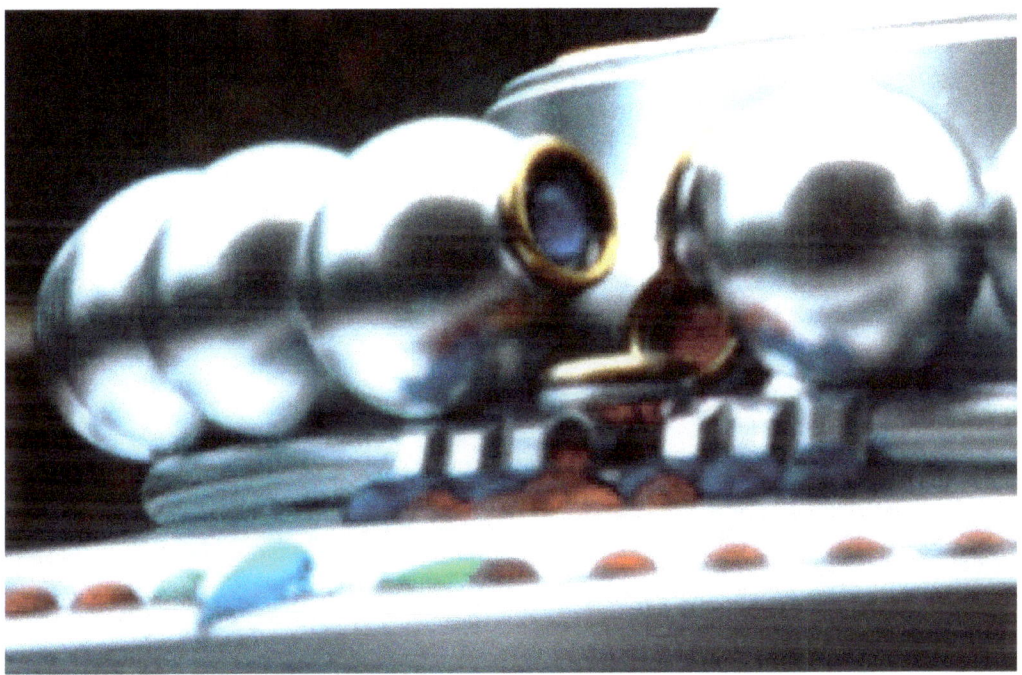

Figure 29 - Details of colourful crystals and lenses in Photo #808.

Large red and blue magnifying lenses appear to be inside the golden rings. It cannot be seen whether the sphere on the right has another lens, but checking night pictures of this WCUFO indicates there is another lens there, possibly a green one (see Figure 30). Red, blue and green are the three primary colours of light which could be significant, but who knows what the significance is here?

Below the spheres, five or six truncated pyramidal or hexagon pieces are in alignment on a base ring in two sets of three — a red crystal nestles inside one middle piece of these six.

They are Here

Finally, the two spheres with golden rings seem extruded towards the rings. It can conclusively be said that these are not Christmas tree balls with golden rings glued on them as some sceptics have naively claimed. Although some Christmas tree balls have extrusions with caps to take a thin wire, the extrusion would be tough to accomplish on a delicate little 4 cm diameter reflecting Christmas tree ball; and that is if this WCUFO were a scale model, and we know it is not.

The top tier of this WCUFO shows curious red, and gold/yellow, reflections (in coloured versions of this photo), on the upper central three spheres (see Figure 31). On the upper central sphere, these gold/yellow reflections are of the golden projections beneath, between the spheres on the second row. The top two red reflections on the two upper deck recessed spheres are from two of the four red crystals on the top platform. A tiny bit of them is visible on the edge of this platform. A line traced from each sphere centre, crossing through the reflection of the red crystals, points to the crystal on the upper platform. They are not visible on any of the other WCUFO photos, except the night shot from above the WCUFO.

Figure 30 - WCUFO at night. A red-reduced photo #873 reveals the top platform red crystals, and blue crystals on the base looking black.

Also seen in the WCUFO night shot are the notches on the top edge of the top tier further recessed towards the centre of the craft just where these recessed spheres are; these deep recesses appear to be red in the night photo despite the general lack of colour. The red remains even when the picture is radically "red-reduced" in Photoshop, (Figure 30 top tier). So it appears that these two red reflections in this daytime photo #808 are lights or crystals set within these upper-tier recesses in front of the recessed spheres.

Embossed and engraved details

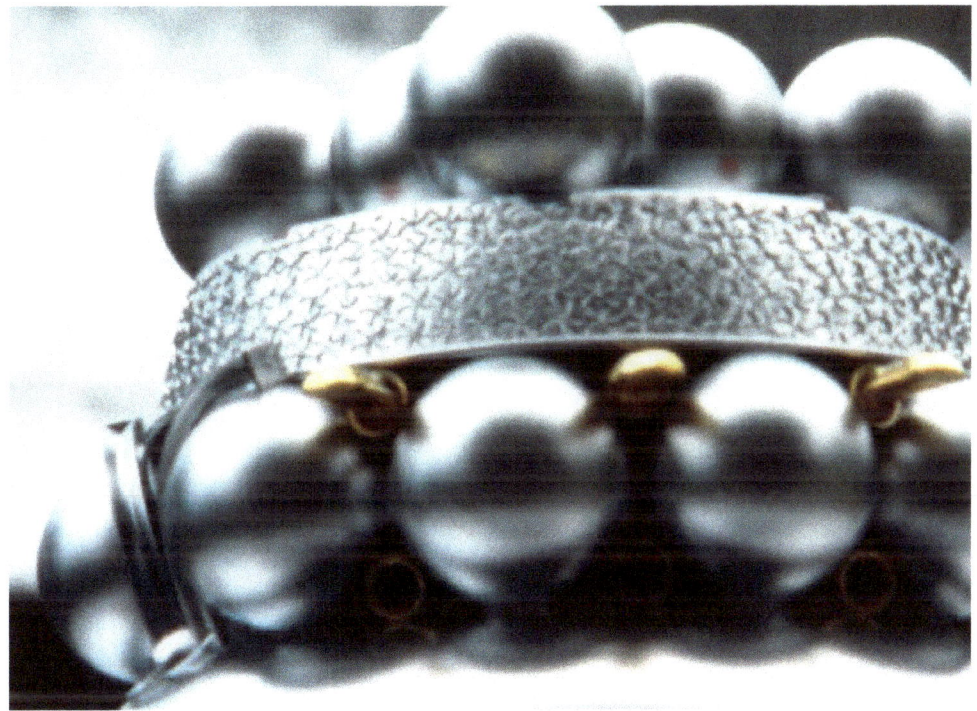

Figure 31 - WCUFO top tier details showing embossed or engraved stars.

The top tier of the WCUFO also has an intriguing pattern of star-like embossings or engravings. On the left, two "V" arcs end in a little rectangular plate. Better detail in Figure 32 shows this plate with a cross in it. If this WCUFO were a scale model made from a 55 cm trash-can lid, this plate with its cross would measure 7.5 mm by 5 mm. It is not impossible to engrave such a tiny detail, but it is a little complicated and completely unnecessary. Meier had no motive to create such unnecessary detail that no one noticed or mentioned for over 30 years.

Some sceptics claim this WCUFO top tier is an available plastic flowerpot tray shown in Figure 33. The two are so remarkably similar that it is easy to wonder if Meier did make a model as sceptics have claimed. Conundrums like this are typical of the Billy Meier case evidence. At first glance, Meier looks suspect of falsifying his evidence. However, upon a full and proper investigation, it is conclusively shown that he is not. Compare the objects in Figures 31 and 33.

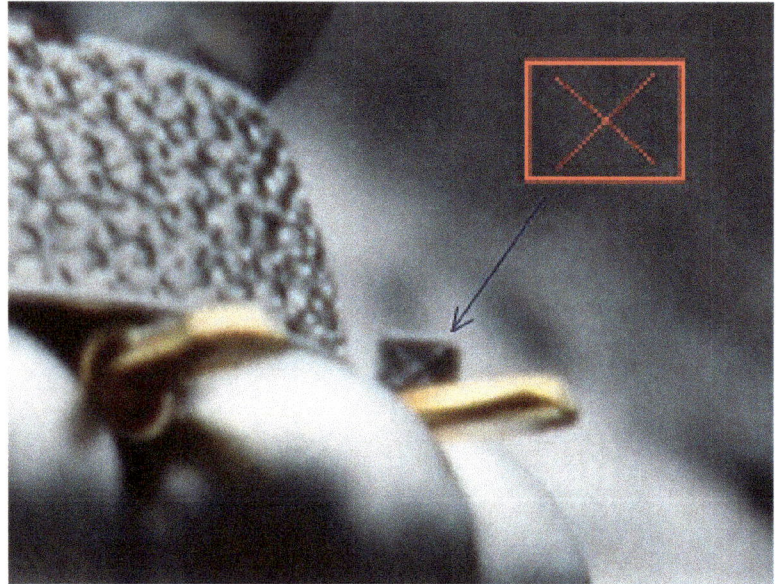

Figure 32 -   Details on the little plate at the end of the "V" shaped arcs

At first glance, they look the same shape, size and pattern. Despite the many similarities, however, they are not the same. First is the difference in ratio between the pot tray width and height. In Meier's photo, the ratio is 5.6 (width divided by height), but the flowerpot tray ratio is only 4.0. The plastic tray has more height or depth than the component in Meier's photo. These trays are available in Europe in different sizes, but no one has ever come forward with one the same size as Meier's WCUFO upper portion. Could Meier have cut the plastic object reducing its height? It has a remarkably smooth edge, and why not just use what was available uncut?

Figure 33 - Plastic flower pot tray with embossed stars presented by sceptics as "evidence" that Meier made his WCUFO with readily available materials.

Counting the number of embossed stars from top to bottom on the plastic tray reveals no more than four, but Meier's photo shows five. So rather than cutting the plastic tray, Meier would have to have added extra stars, and make them smaller. That is not possible for him. So Meier changing the shape is highly unlikely.

It does not end here; however, the stars themselves appear different. In the WCUFO (Figure 31) the star-like motifs appear possibly engraved or embossed, relief motifs as on the flowerpot tray, where they are all clearly in relief. On the top tier just left of centre the star-like motifs appear punched into the top tier face. It is not clear, however, and the WCUFO star motifs can read ambiguously either as embossed or engraved motifs, while the flowerpot tray motifs always appear embossed. Should not the upper edges of the WCUFO stars, where their background is bright (Figure 31 top tier detail, upper left of centre) also be bright from reflecting the overhead light if they are embossed or in relief? They are dark, lacking bright highlights, which supports the notion of engraved images with light unable to enter into their indents, leaving only their darkened shadow. Alternatively, the very bright highlights might be burned out by the bright skylight leaving the embossed star shapes in shadow. Also, on the upper right-hand side of the top tier (Figure 32), the stars can appear darker at the top and lighter on the bottom. Of course, again, this could suggest engravings, not embossing. These stars are like an optical illusion, challenging to read. The flowerpot stars give the opposite effect; their upper half is lighter because being in relief, they catch the light (Figure 33). The far left of this WCUFO upper part, however, can suggest the motifs are embossed or engraved. Which is it? Are these star motifs engraved or embossed on the WCUFO? We are unsure. It is a curious situation. What is your opinion regarding these "stars"?

It must also not be forgotten that the WCUFO is proven to be over 3 m in diameter, meaning this top component would have to be one world record-breaking enormous plastic tray. The top tier is approximately 1/4 the diameter of the WCUFO, which here is 3.5 m, making this upper tier about 875 mm in diameter. Contact us if you find one of these plastic flower trays that big at the local hardware store. Even today, the largest modern plastic flowerpot trays online are around 457 mm in diameter. In the 1980s sizes generally were within 254 mm to 304 mm. So as far as size is concerned, this WCUFO upper tier cannot possibly be a plastic flowerpot tray; and yet the similarities intrigue.

At this juncture, before concluding this part, Meier's explanation for this similarity should be given, and you, the reader, can then decide. Meier says that once he pointed it out to his Plejaren contacts they told him that they gave the basic WCUFO design to scientists in Europe, it seems just before the advent of WW II. Eventually, domestic industries received certain parts of the design plans for commercial use. Hence the WCUFO upper component that looks like an upturned flowerpot tray, and its base that looks like a trashcan lid, were both somehow given to domestic industry for production and commercialisation.

Alternatively, perhaps designers back then just saw these images in industry sketches somewhere and subconsciously brought them forward in their designs. Derren Brown has shown how designers are easily influenced by subliminal imagery to produce their designs ("...tricks advertisers with subliminal messaging" *YouTube*).

Billy Meier's Courtyard WCUFO: A Detailed Analysis

Regarding these WCUFO components, according to Meier, the Plejaren commander Ptaah in Contact Notes 254 and 259 explains the similarities of various WCUFO parts, 17 in total, to everyday household items. Here is the explanatory extract from Contact report 259, of 25 February 1997 between Billy Meier and Ptaah:

Figure 34 - Drawing of Eduard Albert Meier (Billy Meier) by Remington Robinson and a Ptaah self-portrait.

Ptaah

4. *My assumption was correct, that the utilization of the plan, in regard to this flying device, found wider-ranging interest than that merely futuristic forms of container lids were manufactured from them, as I communicated to you at our conversation on November 28th, 1995.*

5. *Our very extensive clarifications have revealed that the flying-device plans were variously divided up and were used for further manufacturing, as, for example, for various forms of decoration and holding devices for various purposes.*

6. *Also, exposed bolts for furniture shelves were created out of certain parts of the plans, as well as roller bearings from other parts of the*

Ptaah

4. *Meine Annahme war richtig, dass die Planverwendung in bezug auf dieses Fluggerät weiterreichende Interessen fand als nur gerade die, dass daraus futuristische Behälterabdeckformen angefertigt wurden, wie ich dir bei unserem Gespräch am 28.11.1995 mitteilte.*

5. *Unsere sehr weitreichenden Abklärungen ergaben, dass die Fluggerätpläne verschiedentlich aufgeteilt und für weitere Anfertigungen benutzt wurden, wie z.B. für verschiedene Schmuckformen und Halterungen für verschiedene Zwecke.*

6. *Aus bestimmten Planteilen wurden auch Aufliegebolzen für Möbeltablare geschaffen sowie aus anderen*

They are Here

plans.

7. A flower pot base came about from another part of the plans, which, in its outer edge part, is very precisely configured in accord with the plan sketch, consequently, therefore, even the surrounding pertinent special elements for the screening of the visibility of the flying body was carried over, and indeed so precisely and correctly that no difference exists between the plan sketches and the terrestrially created product.

8. However these are not the only terrestrially manufactured products from the flying-device plans, because we could determine a total of 17 different objects for different purposes which were worked out from the plans.

9. We have gained no knowledge as to why that happened, if you exclude the fact that the flying-device plans of that time were ripped out of their entirety and were distributed in several European countries as well as in America and Japan where they found utilization for the planning and manufacturing of the most varied products which, in part, are still produced and utilized on Earth today.

Planteilen Rollenlager.

7. Aus einem andern Planteil entstand eine Gewächstopfunterlage, die im äusseren Randteil figürlich sehr genau den Planzeichnungen entspricht, folglich also sogar die rundum angebrachten Spezialelemente für die Ausblendung der Sichtbarkeit des Flugkörpers übernommen wurden, und zwar derart genau und korrekt, dass kein Unterschied besteht zwischen den Planzeichnungen und dem irdisch erstellten Produkt.

8. Dies sind jedoch nicht die einzigen irdisch hergestellten Produkte aus den Fluggerätplänen, denn gesamthaft konnten wir 17 verschiedene Gegenstände verschiedener Verwendungszwecke eruieren, die aus den Plänen herausgearbeitet wurden.

9. Warum das geschah, darüber haben wir keine Erkenntnisse gewonnen, wenn man davon absieht, dass die damaligen Fluggerätpläne aus ihrer Gesamtheit gerissen und in mehrere Länder Europas als auch in Amerika und Japan verteilt wurden, wo sie dann für die Planung und Herstellung der verschiedensten Produkte Verwendung fanden, die teilweise noch heute auf der Erde hergestellt und verwendet werden.

This WCUFO parts issue is an excellent and typical example of the challenge facing investigators into the Billy Meier case. At first glance, the topic focused on seems a purposeful assembly of several details suggesting the case is false. Still, upon performing detailed analysis, the point under investigation turns out to be factual as discussed earlier in this book's *Introduction - A Worldwide UFO Controversy* (pages 1 ~ 11). We call this apparent dichotomy "the exit door" or a way out for people who cannot

Billy Meier's Courtyard WCUFO: A Detailed Analysis

handle the information. An *exit door* is beneficial for people that could otherwise become severely impacted to the point of mental imbalance by knowing the Meier case – or any UFO case – is real. An *exit door* enables such people, notably those with fixed religious or philosophical beliefs, to get out without being psychologically harmed. Their *exit door* could be to simply conclude, "I have my doubts," "I'm not so sure about that," or "There is no proof." Alternatively, as is quite commonly said in simplistic sceptic communities, "If it looks fake, it must be fake." A growing number of people, however, find that although something may look fake at first, upon proper analysis, it proves factual:

> ***Things are not always what they seem; the first appearance deceives many; the intelligence of a few perceives what has been carefully hidden.***
>
> *— Phaedrus*

For additional interest here is part of Contact Report #442, of 10 February 2007, in which Ptaah eventually gave some measurements of the WCUFO hovering above Meier's parking lot (Frehner email, German trans. Devine):

Billy

The same goes for my side. - Some time ago, Professor James Deardorff, from the United States, wanted to know some things regarding the cake ship that was landed in our car park between the shed and the house. The history - in regard to the striking similarity of the ship's hull with a barrel cover shape, that is to say, with a drum lid, and the emergence of those storage drums which we discovered on our property - is listed in the 254th contact report. Because of that, Christian Frehner had to take various photos to specify how large the ship was and the distance between me with the camera and

Billy

Das gilt auch meinerseits. – Vor geraumer Zeit wollte Professor James Deardorff aus den USA einiges wissen bezüglich des Tortenschiffes, das auf unserem Parkplatz zwischen der Remise und dem Wohnhaus gelandet wurde. Die Geschichte in bezug auf die frappierende Gleichheit des Rumpfes des Tortenschiffes mit einer Fassabdeckform resp. mit einem Fassdeckel und dessen Entstehung von sich in unserem Besitz befindenden Lagerfässern ist im 254. Kontaktbericht aufgeführt. Christian Frehner musste deswegen verschiedene Photos machen und angeben, wie gross das Schiff und die Distanz zwischen mir mit dem

They are Here

the cake ship. For my part, I know that the diameter was approximately 3.50 meters, which I also told Christian. But at the same time, can you now also provide for me the exact measure, as well as how many people can be accommodated in this small ship?

Ptaah

3. This is no secret, and I know the precise data:

4. The lowermost outermost diameter, with the flat edge, was 3.52 meters, however, the outermost upper outer rim diameter was 3.20 meters.

5. The entire outer edge structure, on which the swinging-wave accumulators were attached, was 37.6 centimeters, while also with this, the measurement of the bottom ring to the level of the swinging-wave accumulators was 32 centimeters.

6. And room in the flying device was designed for a person in a sitting position, but in an emergency three persons could find room in a crowded manner.

7. These types of flying devices were not suitable for the Earth's atmosphere, for which reason, after a brief time of service, they were withdrawn from terrestrial space.

Billy

That was not known to me. With the swinging-wave accumulators, the silvery spheres are probably meant,

Photoapparat und dem Tortenschiff war. Meinerseits weiss ich, dass der Durchmesser rund 3,50 Meter beträgt, was ich Christian auch sagte. Kannst du mir dazu nun aber noch das genaue Mass angeben, aber auch wie viele Personen in diesem kleinen Schiff Platz finden?

Ptaah

3. Das ist kein Geheimnis, und die Daten sind mir genau bekannt:

4. Der unterste äusserste Durchmesser mit dem flachen Rand betrug 3 Meter und 52 Zentimeter, der äusserste obere Aussenranddurchmesser jedoch 3 Meter und 20 Zentimeter.

5. Der gesamte äussere Randaufbau, auf dem die Schwingungskumulatoren angebracht waren, betrug 37,6 Zentimeter, während das Mass vom unteren Randring bis zur Ebene der Schwingungskumulatoren 32 Zentimeter betrug, wie auch bei diesen.

6. Und ausgelegt war der Platz im Fluggerät für eine Person in sitzender Stellung, wobei aber notfalls drei Personen in gedrängter Weise Platz finden konnten.

7. Diese Art Fluggeräte haben sich für die irdische Atmosphäre jedoch nicht geeignet, weshalb sie schon nach kurzer Einsatzzeit wieder aus dem irdischen Raum abgezogen wurden.

Billy

Das war mir nicht bekannt. Mit den Schwingungskumulatoren sind wohl die silbrigen Kugeln gemeint, nehme

I suppose, or?

Ptaah

8. That is right.

9. Their diameter was 32 centimeters, as I already said.

Billy

Thanks for the data, which Christian can probably use.

ich an, oder?

Ptaah

8. Das ist richtig.

9. Deren Durchmesser betrug 32 Zentimeter, wie ich bereits sagte.

Billy

Danke für die Daten, die Christian vielleicht gebrauchen kann.

In this Contact Report #442, Ptaah gives specific measurements of this WCUFO. It also states that the WCUFO holds one pilot or three in a crowded manner in an emergency. These WCUFOs are no longer here. Why?

Meier says the Plejaren removed their Earth bases in 1995 and closed down their operations or stopover places here. Presumably, they had completed their planned research or work here and no longer need to fly in our skies with the WCUFOs.

Perhaps the WCUFOs withdrew due to corrosion caused by our atmosphere. Maybe with a cover, like other beamships, they might be more suitable and last longer.

Interestingly Ptaah refers to the spheres as "swinging-wave accumulators." Perhaps we might call them something akin to frequency or resonance accumulators. Is he referring to gravitational waves that could be modified to produce anti-gravity waves?

We should also note here at the end of this section that both Billy Meier and Christian Frehner think that Meier used his 42 mm Olympus 35 ECR to take these WCUFO courtyard photos. Still, our calculations reveal him using a 55 mm lens, like his Ricoh camera with its 55 mm lens that was invariably too difficult for him to use with only one hand. It is possible, however, that he tried just this one time immediately outside his house, possibly making use of a tripod or some local support, and hereafter returned to using his Olympus 35 ECR camera exclusively. Photo #808, for example, shows the nearby WCUFO in sharper focus than the distance which is impossible for Meiers Olympus 35 ECR camera that is stuck on infinity focus.

Finally, Figure 35 diagram illustrates our calculation of the WCUFO's movement as it hovered above Meier's parking lot. Using the eleven photos available (Figure 8), and estimating the positions based on the 3.5 m size already arrived at for pictures #799, #800 and #808, and the timing, size, and location on each photo, we produced the WCUFO flight pattern in Figure 35.

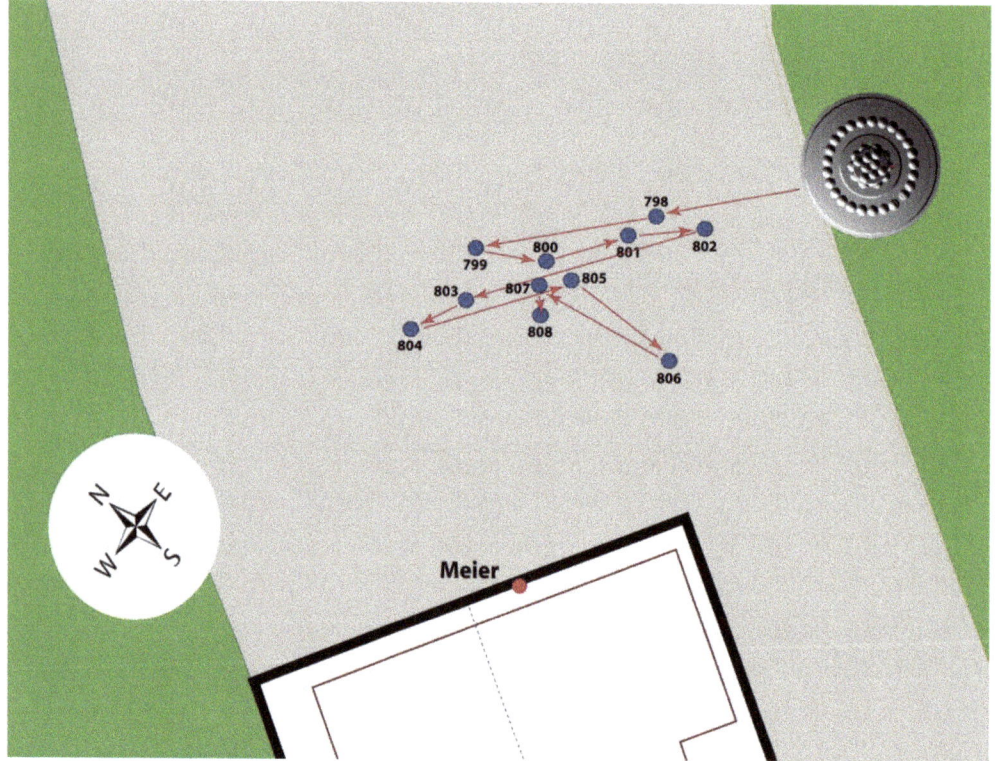

Figure 35 - WCUFO's movement above Meier's courtyard plotted from his 11 sequential photos. Dots indicate the approximate WCUFO centre in each photo. Meier is proximal to the carriage house wall. The 3.5-metre diameter WCUFO shown assumes a southeast arrival.

Finally, the most recent tests done by Christian Frehner with an Olympus ECR camera at the parking lot confirmed that Meier did not use this camera, but a 55 mm focal length one. So Meier used the reported Ricoh camera, not his old Olympus. However, it matters not which camera he used, the results in this analysis would be the same, because:

- We know by checking the alignments of visible details in the background home image that Meier was located very close to the carriage house wall. It does not depend on the type of camera he used.

- We know the distance to the WCUFO central sphere from the carriage house is more than six metres, whichever camera he used.

- We measured the WCUFO angle (or semi-angle) of view by checking alignments of the central axis and the WCUFO edge pointing to background details in the image. We used two methods, one dependent on the type of camera and the other not dependent on the camera type, to measure the WCUFO angle (or semi-angle) of view by checking alignments of the central axis and the WCUFO edge pointing to background details in the image. (Zahi *Analysis of the Wedding Cake UFO...* Annex B).

- The angle of view together with the camera's distance from the WCUFO show an object of 3 m to 3.5 m in diameter, not a small model

- The WCUFO photos at Meier's parking lot reveal a camera lens characteristic known as pincushion image distortion showing the WCUFO asymmetrical. So these are full-frame photos compatible with the Ricoh camera lens distortion. Considering a photo taken with the Olympus 35 ECR camera cropped to account for the angle of view, the distorted pincushion effect part would not appear because it would be in the portion of the photo cropped off. The WCUFO image would also not extend to the edges of the cropped photo, but be near the centre where this pincushion distortion is absent in both cameras.

Meier may well have used his Olympus camera in other pictures of the WCUFO, but not for the photos taken at his parking lot. We consider the issue of which camera Meier used here an option for possible future research.

WCUFO among the Treetops: An Analysis

Introduction and key findings

On 3 April 1981, Meier took pictures of another 3.5-metre WCUFO, referred to here as "the WCUFO", flying close among some treetops. Meanwhile, Meier's records say he photographed it among the treetops while standing outside and on the top of a beamship. Thus two spacecraft were involved, both apparently near treetop level. All photos shot among these treetops, some of which are the subject of this investigative analysis, show the WCUFO. Interestingly, Meier claims he never got inside a WCUFO. We also note that while *Through Space and Time* says Meier was "sitting on a second ship" (pages 24 ~ 26 and 44 ~ 46) we recently confirmed through Christian Frehner that "sitting" there meant "situated on" not sitting on his backside; he was *standing* on it.

Eventual conclusions:

- The sphere reflections in the WCUFO photos show a surrounding forest (#834, #838 and others). Unable to precisely quantify the WCUFO size from these photos, we could only estimate its size. Still, it is undoubtedly not a small scale model. A forest of trees is shown between the camera on-board the beamship and the WCUFO. So if the WCUFO had been a small model, these trees would appear much more prominent in the reflected images. The only possible conclusion is, the WCUFO is not a scale model photographed close to the camera but a distant object above the middle of a forest.

- Although it is next to impossible to discern visually since its image is so tiny and blurred, we locate within the sphere reflections, the 7-metre beamship that Meier claims he was standing on while taking the photos. Of course, if the original photographs were available to digitise in better resolution, a better view of the forest and beamship would be possible.

- In photos #834 and #838, various forest treetops are visible at different distances, especially when viewed in 3D. An excellent 3D image of the WCUFO hovering in front of a nearby tree is in our book *Researching a Real UFO*, and the book includes instructions on how to construct a stereoscope to view the stereo pair of images.

- Meier may have used either his 42 mm focal length Olympus ECR Camera or his 55 mm Ricoh camera that he used to shoot the WCUFO photos in his parking lot. It does not matter which, because in either case, the result is the same.

WCUFO among the Treetops: An Analysis

Photo #838 analysis

Photo #838, taken by Meier on 3 April 1981, shows the WCUFO flying close to the treetops (Figure 36). As mentioned, Meier was standing on the outside of a 7-metre diameter beamship 20 to 40 metres above ground taking the WCUFO pictures. So, there are two crafts involved, both apparently near treetop level. Photo #838 is one of a series of several photos of this 3.5-metre diameter WCUFO.

Figure 36 - Photo #838 taken at treetop level. The WCUFO is proximal to a tree.

Based on the WCUFO size in this photo and the characteristics of Mcier's camera, calculations show that if he used his Ricoh camera, this is a 3.5-metre WCUFO at 25.4 m away from him, or 19.4 m away from him if he used his Olympus camera. If it was a scale model of 55 cm in diameter, it should be just three to four metres from the camera.

They are Here

A UFO scale model at three to four metres from the camera, with a tree in the middle as in photo #838, means the WCUFO must be so close to this tree that the tree's reflection on its spheres would be way too big for what the picture reveals.

WCUFO: computer model simulation

Using the tool "Blender", we made a 3D model of this WCUFO, based on the proportions for the 3.5-metre WCUFO measured from several photos.

The figure below shows our modelled 3D representation. The camera is located onboard the model 7-metre beamship. Meier says he was standing on this beamship approximately 20 to 40 metres above ground and there took picture #838 of the WCUFO, at 25.4 m away. Using either a 3.5-metre or 3.0-metre WCUFO similar image results arise in these sphere reflections. We show, momentarily, that this is not the case when using a small-scale 55 cm model WCUFO.

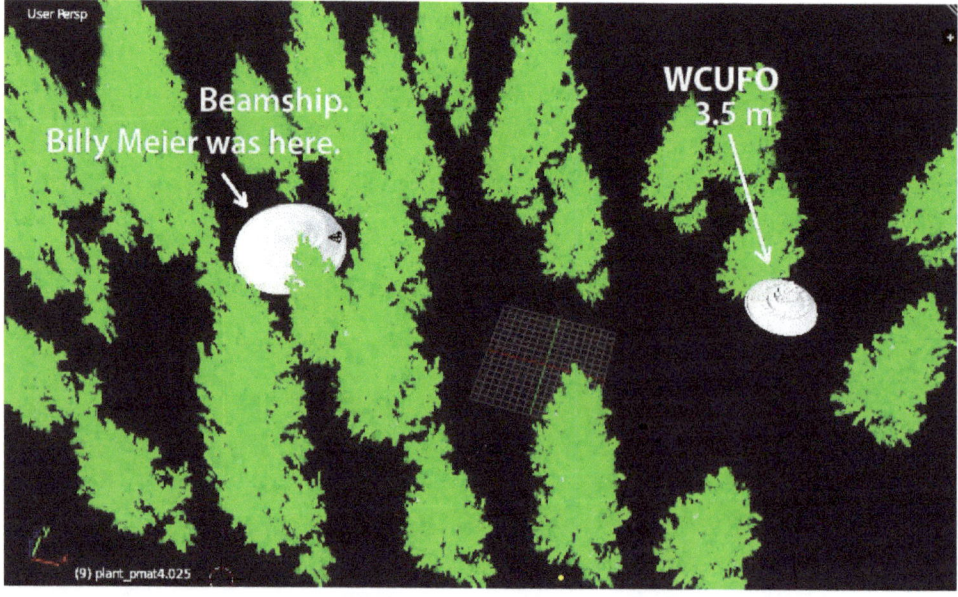

Figure 37 - A computer-generated model of the 7-metre beamship and 3.5-metre WCUFO and the forest in photo #838.

A few stylised trees are included as a reference. The beamship is behind a tree that often partially interferes with its view of the WCUFO (Figure 42).

Using Blender software rendered the beamship model image viewed from the camera. The model camera used had a 55 mm focal length, the same as Meier's Ricoh camera. If Meier used the Olympus, the resulting image would be the same in apparent size, but we need to locate the WCUFO a bit closer to the camera.

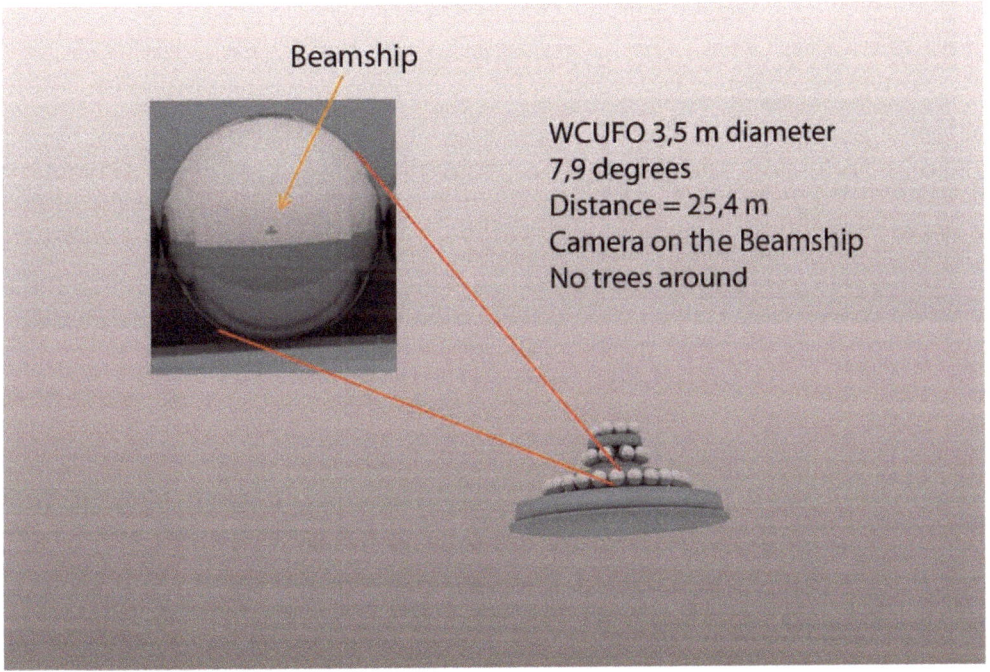

Figure 38 - Sphere reflection on the computer-modelled WCUFO showing the tiny 7-metre beamship image. (European numeral system of commas for decimal points.)

The first rendered picture is processed without trees to show the size of the beamship reflected on the WCUFO spheres. Figure 38 inset shows the result: sphere magnification reveals the reflected beamship image is just a tiny dot at the very centre of the sphere. However, the reflecting metal sphere also reflects the dark surroundings with all of its trees, making it extremely difficult to see the beamship in Meier's photograph. See Figure 39 in which the repeated process now renders the scene with stylised trees.

They are Here

Figure 39 - Reflection on the 3.5-metre WCUFO with a simplified rendition of the surrounding forest. (Assuming a Ricoh camera.)

Even though this is a simplified and more lucid computer model, it remains challenging to find the beamship and camera's location, reflected in any of the WCUFO spheres. So it would be extraordinarily difficult to see in Meier's photo. However, as dictated by sphere reflection rules, it is located precisely in the centre of the sphere image, partly or mainly behind the trees.

Momentarily this rendered image (Figure 39) is shown similar to the real picture reflections (Figure 41). While we included just a few trees, with several trees of the surrounding forest, the rendered image would be very similar to Meier's photo. Conducting the same tests with a 42 mm focal length Olympus camera, the results would be very similar, given the long distance to the WCUFO.

The trees, with stylised rectangular leaves replacing conifer needles, only roughly represent the forest in Meier's photo; their only purpose is to approximate the size of the reflected trees. Also, because it is unnecessary for our purposes, the WCUFO model here is not a detailed representation of the real WCUFO. Still, the spheres necessarily do occupy the same positions, and details are proportionally the same as in the WCUFOs photographed by Meier.

Because some Meier case sceptics claim this WCUFO is a scale model made from a trash-can lid of approximately 55 cm diameter, we render another WCUFO model of 0.5 m diameter located 4 m away and compare the results. See Figure 40, which omits the beamship.

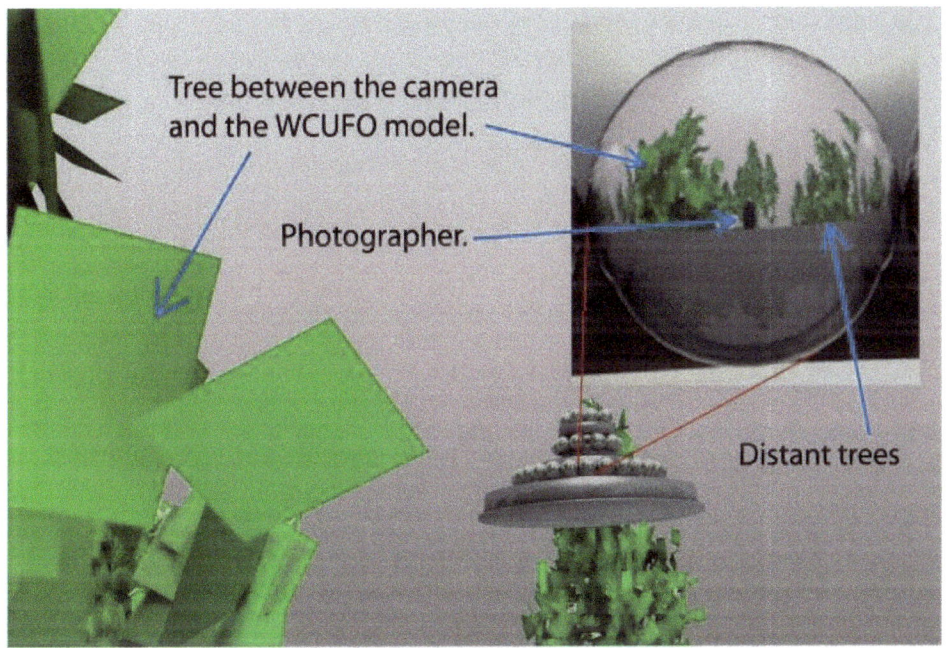

Figure 40 - Reflection on a 0.55-metre scale model showing nearby trees and the dark rectangle at the sphere's centre representing the photographer and the craft he was on. (Assuming a Ricoh camera.)

The results differ significantly from those of Figure 39. The same render using an Olympus camera would produce a slightly bigger tree reflection because the UFO model would be even closer to the camera. The trees look way too large due to the WCUFO scale model's proximity to the trees and camera. Similarly, a photographer managing to climb such a high tree and taking such a photo would be visible in the sphere's reflection due to his proximity to the model. The reflections in the real photo (shown momentarily) look nothing like the configuration obtained with a

They are Here

0.55-metre scale model, revealing that Meier did not use such a model.

Sphere reflections: analysis of photos #838 and #834

Figure 41 shows an image-processing of photo #838. The image of one central sphere is magnified and increased in contrast and saturation to aid viewing. As shown in this photo, there is a forest around the WCUFO. For a better 3D view of the trees around this UFO, stereo images of this photo's sphere reflections are in our book *Researching a Real UFO*.

Figure 41 -   Reflections on actual photo #838 showing a forest around the ship.

Figure 42 shows that photo #834 also reflects the surrounding trees. Some trees are visible even though the images are not very clear. The beamship Meier stood on while taking this picture must be at the centre of the sphere image. Invisible behind the trees, it is neither atop the highest trees nor at the bottom of the forest. Meier says he was approximately 40 metres above the ground on the top of the beamship. The photo shows the WCUFO near the top of these trees, and only a very narrow-angle WCUFO base is visible. So the camera is perhaps no more than a metre or two below the WCUFO base depending on how far away it is, and this establishes the fact that Meier's camera must be situated somewhere in the upper range of the trees.

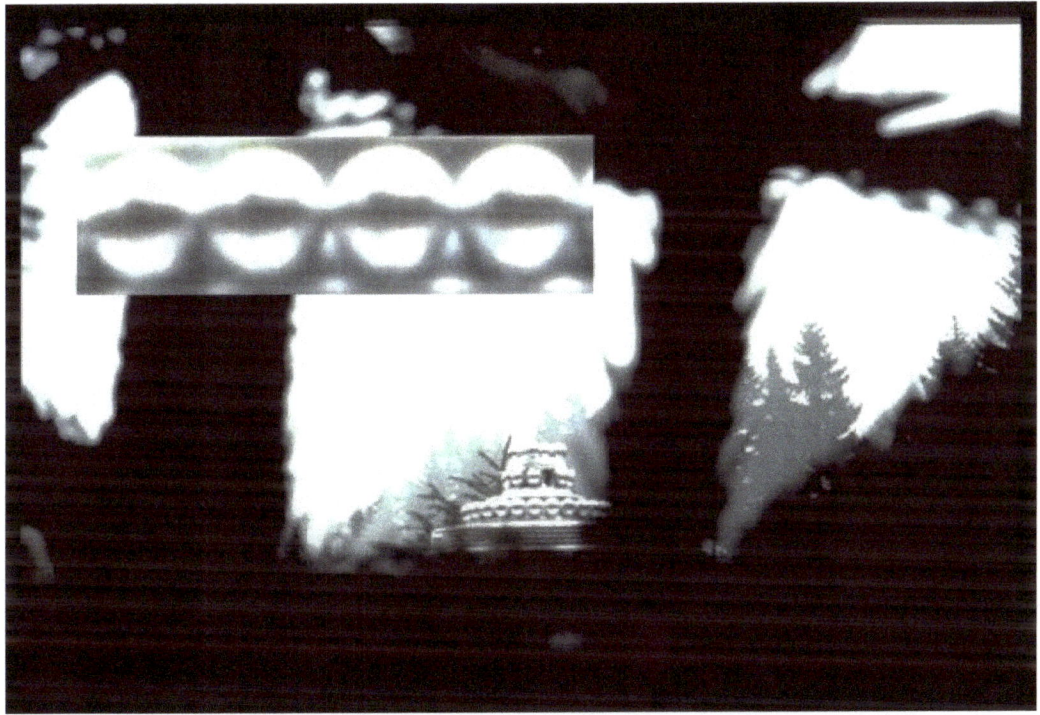

Figure 42 - Photo #834. The WCUFO shot through or behind a tree. Sphere images show small, low elevation and therefore distant images of the surrounding forest.

Returning to the reflected image in Figure 41, #834, the upper part of the reflected image shows a clear sky, with no nearby trees. The middle section shows a band of trees from the surrounding forest, and in the lower part of the sphere's image, the bright sky reflects from the WCUFO base tier. A reddish area in the middle could be the reflection of the red crystals on the outer edge of the WCUFO base tier. The orangish colour on the top left edge of the sphere might be caused by a sunlit cloud or the sun just partially

They are Here

breaking through a cloud. Also visible are different shades of green in the forest probably showing trees at various distances from the WCUFO.

Comparing Figure 42 with Figures 40 and 39, the sphere reflections in Figure 42 are similar only to those in the simulation of the distant WCUFO. The craft cannot have been a very close small model since no close trees project upwards on the spheres; a small model would have to be close to the trees resulting in large branches or pine needles reflected in the image. This fact alone leads inescapably to the conclusion that this WCUFO is not a small model but a large craft.

In this analysis, however, the precise size of this WCUFO is uncertain. Estimation of its size is, however, possible by studying other pictures, and the trees behind it. Again, it is not a small-scale model photographed in the forest since there is no upward extending reflection of the nearby tree between the camera and the WCUFO. Furthermore, for a scale model at just four metres from the camera, the photographer and the nearby tree would be highly visible in the sphere reflections.

Our calculations based on the branch sizes showed the nearest tree, the closest one in Figure 36, is around 3.3 m away from the camera, if the Ricoh, or 2.5 m using the Olympus. So a 55 cm model UFO would be approximately 0.70 metres behind the tree in front of it. The model being proximal to or inside the tree, but the reflections do not show this.

In photo #834 (Figure 42), again, as seen in the sphere reflections, a forest is visible all around the WCUFO. Magnifying the image and increasing the contrast and saturation, the trees are fuzzy but visible. This treetop is probably the upper portion of the tree whose thick trunk or stem is seen prominently nearby on the left side of Figure 42. The off-vertical "barbed" shape on the right is a proximal out-of-focus limb extending upwards from the tree, as mentioned above. In photo #834, Meier stands on and takes pictures from the 7-metre beamship hidden from view behind the tree, as in #838.

The next Figure, 43, details the reflected image in this picture.

WCUFO among the Treetops: An Analysis

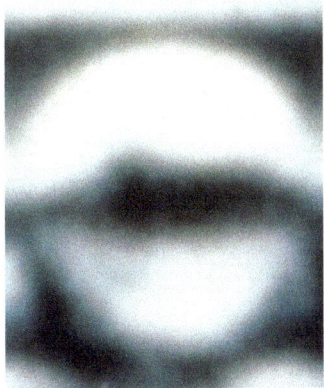

Figure 43 - From photo #834. Enlarged view of a sphere with centrally oriented beamship hidden by the tree.

Finally, the estimated size of this WCUFO is around 3.0 m to 3.5 m, based on the distances to nearby trees. Further details of this estimation are available in the 74-page report: *Analysis of the Wedding Cake UFO* (Zahi).

Daylight Full-View Full-Size WCUFO

Big tree or little tree, with proximal WCUFO?

We demonstrated that the pictures of the WCUFO hovering above Billy Meier's parking area show a 3.0- to 3.5-metre diameter WCUFO, the size of a small car. Also shown is the fact that the WCUFO flying above the treetops could not be a small scale model but is a sizeable object. There are, however, other ways to know how big this WCUFO is. Once the size of a nearby tree is somewhat accurately estimated, and the WCUFO's presence at that tree established, we know again roughly how big the WCUFO is.

Some sceptics claim that Meier used two types of tricks to fake his photos: either a little model with small trees proximal to it; or a "false perspective" trick of taking a photo of a nearby model with a big tree several metres behind in the background. The following figure shows two photos that some sceptics think to fall into each of these categories.

Sceptics have made photo tests of a distant little tree close to a distant scale model WCUFO (simulating photo #841 in Figure 44, top image). It is evident from zooming their test images that the little tree was cut with clippers to make it appear similar to a real tree. They also tested a significantly sized real distant tree, with a scale model in front of the camera, the false perspective trick, to get something like photo #844 in Figure 44, bottom image. Noticeably, however, by enhancing details of the branches in Meier's photo #844 and #841, it is quite clear that this is a real and enormous tree, rendering the small tree hypothesis nonsense. Could this photo #841, however, be a trick of false perspective?

They are Here

Figure 44 -   Photo #841 (top) shows a distant tree close to the WCUFO. Is this a little tree and tiny UFO model, both far away from the camera? Photo #844 (bottom) tree branch details confirm the WCUFO is in front of a big tree.

Establishing this WCUFO's proximity to a big tree would confirm it a significantly sized object. If the tree is not a bonsai or little tree, but of significant size and the WCUFO is right up close to it, the tree size enables estimation of the WCUFO size. Four ways demonstrate that in these photos this WCUFO measures around 3.5 metres in diameter: These ways are discussed under the following four clue headings:

Clue 1. by photo sequence

      2. by colour reflections on the WCUFO undercarriage

      3. by sphere reflections

      4. by WCUFO shadows cast upon the tree.

The following clued explanations demonstrate each way.

They are Here

Clue 1: The photo sequence.

Figure 45 -   Photo sequence from #840 to #844. 10 April 1981.

In this photo sequence (Figure 45), taken somewhere in the countryside between Girenbad and Hinwil, on 3 April 1981, between 1:15 pm and 1:35 pm, a stationary WCUFO is proximal to a young but substantial tree. Judging by the grass and distant objects, like the little bush indicated by the red arrow, Meier is walking up a little hill towards the WCUFO and tree as he shoots the sequence. When the sequence ends at picture #844, Meier, standing, has now moved left. So the WCUFO is now in front of only the right-hand side of the tree. In the bottom far right is a distant forest only visible in the last photo. Most of these images are courtesy of the book *Photo-Inventarium* (Meier, pages 110 ~ 113), but two are electronic copies.

It is the same WCUFO in photos #840 - #844, close to the same tree, on the same day, but at different distances. Notice the zoomed WCUFO image in picture #841 shows the same tree as picture #844. The WCUFO might appear to be in a different orientation, but it is stationary; it is Meier moving to the left side of the tree that causes it to show itself now on the right side of the tree.

Since photo 844 shows a big tree, pictures #840 to #843 also display the same big tree close to the WCUFO. The tree is not tiny or a bonsai as some sceptics have naively suggested.

Because the tree and the WCUFO are proximal to each other as pictures #840 to #843 reveal, the only conclusion possible is that this is not a little model close to a little tree. We can only conclude that picture #844 is not the result of a false perspective trick, and this object is of a significant size close to a big tree.

Clue 2: Colour reflections on the WCUFO undercarriage

Figure 46 shows photo #844 minimally enhanced with Photoshop to reveal important details. [6]

The dark green shadows on the WCUFO undercarriage are reflections of the proximal tree of similar darkness and colour. The sun is at the top right, behind the photographer, and casting a considerable shadow over the tree, thereby confirming the big tree's proximity to the WCUFO. If the tree was far away, behind the WCUFO, this dark green reflection could not be on the WCUFO undercarriage, since there would be nothing to reflect, and the left side of its base would appear brighter, like its right side. So this WCUFO is in very close proximity to the tree making it again, of course, a very sizable object.

They are Here

Figure 46 - Enhanced photo #844 shows dark green reflections on the WCUFO undercarriage the same dark colour as the shaded green tree.

Clue 3: Sphere reflections.

Observing the sphere reflections closely, the large size of the tree reflections stands out. If this tree were far away from the WCUFO, it would be impossible to see this reflection so significantly. Figure 47 shows the projected tree reflections, and the left side of each WCUFO sphere reflects the tree as a dark triangular shape (see enlarged inset). All lines traced from the sphere centres reach the top of the tree reflected on the spheres. This property of the sphere reflections indicates the tree is real and close behind the WCUFO; every sphere reflection shows this tree.

Daylight Full-View, Full-Size WCUFO

Figure 47 - Photo #844. Reflections on the left of the WCUFO spheres showing the nearby tree as a dark green triangular shape.

Performing other calculations to position the reflection indicates the tree is at 120 to 140 degrees from a line connecting Meier's camera with the centre of the WCUFO. The guidebook *Researching a Real UFO* presents a method for mapping the local objects surrounding the WCUFO, like the sun, and for finding their horizontal and vertical angles and positions relative to the sphere under analysis (Zahi and Lock "Experiment 8 Mapping the WCUFO's local environs").

So again, we know the tree is close, on the left side, and just a bit behind the WCUFO. If this were a big distant tree, this angle would have to be close to 180 degrees and the tree hardly visible in the reflections. So, finally, again, we see a large object is close to a large tree.

They are Here

Clue 4: WCUFO shadows cast upon the tree.

Figure 48 - Photo #844. The outlined polygon shadow cast on the tree branches by the WCUFO that blocks the sun's rays, indicated by the arrows and confirmed by the bright light reflection on the WCUFO front right edge.

Probably the most crucial clue to knowing whether this tree is far away from the WCUFO or very close to it is the shadows cast by the WCUFO on the tree branches. Fortunately, in picture #844, the sun is shining and producing sharp shadows. The shadow cast is very evident in Figures 48 and 49. The large, bright and dark branches within the delineated red polygon area at the bottom of the tree in Figure 48 indicate where the WCUFO shadow is cast, and the yellow arrows at top right indicate the direction of the sun rays.

This shadow is seen cast by the WCUFO, though to a lesser degree in the reasonably sizeable electronic reproduction of photo #841 (Figure 50). Figure 49, however, is a better zoomed-image of the WCUFO and tree in photo #844. Notice the shadows the

WCUFO casts on the large tree branches and the reflection on the WCUFO undercarriage, just as in #841 (Figure 50), showing it is indeed very close to the tree. It is important to note that the tree has a sturdy trunk (in #841, Figure 50) and sizeable branches. The WCUFO undercarriage is in full view completely refuting some hollow sceptic claims that its undercarriage is never seen in full with the whole craft. The tree's large pine needles demonstrate a sizeable Norway spruce, not a miniature.

Figure 49 - Photo #844. Details of shadows cast on the large tree branches, and reflections on the WCUFO undercarriage.

These four clues inform unequivocally that this WCUFO is a big object, very close to a big tree. The WCUFO diameter, as indicated in photos #840 to #843, is approximately half the height of the tree. According to the *Photo-Inventarium* book description (page 111), this is a young 15-metre tall tree. However, after confirming with Christian Frehner, who confirmed with Meier, we hear the height of this tree was, in reality, seven to eight metres, and that the height printed in *Photo-Inventarium* is an error. From this, we tentatively conclude that this WCUFO has a diameter of approximately 3.5 metres, the same size as the WCUFO Meier photographed in his front yard, and the one he shot at treetop elevation.

Figure 50 - Zoom of Photo #841. More details of another WCUFO shadow cast on the sizeable tree and branches together with the WCUFO undercarriage reflection revealing the very close proximity of the two.

The tree might be a bit smaller or more substantial than this estimate, but it is certainly not a little 1-m high tree as falsely claimed by some sceptics who have not performed sufficient research on this series of photos to arrive at the truth. Photographs #841 and #844 show both a sizeable tall tree and a real large-size WCUFO. Figures 49 and 50 show details of the shadows cast on the large tree by the WCUFO, and reflections on the WCUFO undercarriage. The tree is a Norway Spruce, which is widespread in Switzerland and throughout other parts of Europe.

Following shortly, is an analysis showing the size of a night-time WCUFO could be seven metres or twice the size of this one.

Large WCUFO partly obscured by a large tree in photo #850

One hour later on the same day, 3 April 1981, at 2:33 pm Meier took photo #850. See Figure 51. This photo not only shows a WCUFO close to a large tree, but the WCUFO partly obscured behind it. Behind the tree, as it is, eliminates the false perspective trick, and there is zero evidence of photomontage. Furthermore, enlargements show the WCUFO spheres reflect the tree in front of it excellently.

Figure 51 -   Photo #850. Top: Poster version. Bottom: Enhanced image showing a large WCUFO obscured partly by a large tree in front of it.

Also, as seen in the enhanced version of the photo, Figure 51 bottom, this is another big Norway Spruce, as in photo #844, and perhaps even taller. Noticeably, this Norway Spruce is not the same as the one in Figure 50; the former is more extensive and has a few pinecones on its base, demonstrating again that the WCUFO is indeed a substantially sized flying object.

Below the WCUFO with Trailer and Generator

Using a tractor on 26 March 1981, Meier towed a generator on a flatbed trailer together with his short green travel trailer, used for holding his camera and video equipment to the Swiss countryside (see Figure 52 photo #829). Note the parked travel trailer and generator at the bottom centre of the photograph. This Meier WCUFO photo presents another full view of the WCUFO undercarriage, which here occupies about 61% of the picture width. The photo is a full-frame view with no cropping.

Figure 52 - A full-frame view of a WCUFO with trailer & generator.

Figure 53 (photo #866) gives a more detailed view of the travel trailer and gives another view of the treetop said to have been

They are Here

broken off by Quetzal's beamship earlier. The travel trailer is approximately 1.8 ~ 1.9 m in height to the top of its door.

Figure 53 - Photo #866: Meier's green travel trailer with the treetop piece broken off by a beamship.

Previous Research and Meier's Camera

Previous writings on photo #829 by Professor Deardorff cited Meier taking photo #892 with his 55 mm lens Ricoh SLR camera. We asked Meier in 2009 through Christian Frehner which camera he used, and Meier replied that he took this photo with his old

Olympus 35 ECR camera. Since the event happened 30 years ago, however, and Meier did suffer a considerable loss of memory at one stage, perhaps his memory fails a bit here, as with his WCUFO courtyard photos? So, we now attempt to confirm which camera he did use by analysing essential elements within the photograph and checking them against the cameras' specifications and depth of field data. The analysis definitively informs whether the craft is a small model or a large WCUFO.

On this somewhat overcast day with some sky on the right, and dark trees occupying the left third side of the photo, Meier, near the trees, was probably on a camera stop of f5.6 ~ f11. Looking at the photo details more precisely helps us get closer to which one, and hopefully also which camera he used. The depth of field (DoF) of the photograph is of great help here. Depth of field is the range in which objects are in focus within the photo.

Below are the two cameras that Meier used to take his UFO photographs (Figures 54 and 55). He used the Olympus up until about 1980 after which he apparently and very minimally used the Ricoh SLR with a 55 mm lens; he still preferred his Olympus 35 ECR. (Specifications for both cameras in Annex A). The confusion over which one he used, in this case, arose because it is around this time of 1980 to March 1981 that people thought Meier switched from his Olympus to the Ricoh that someone gave him.

Figure 54 -   Meier's 35 ECR stuck on infinity. (Cameramanuals.org for a manual.)

They are Here

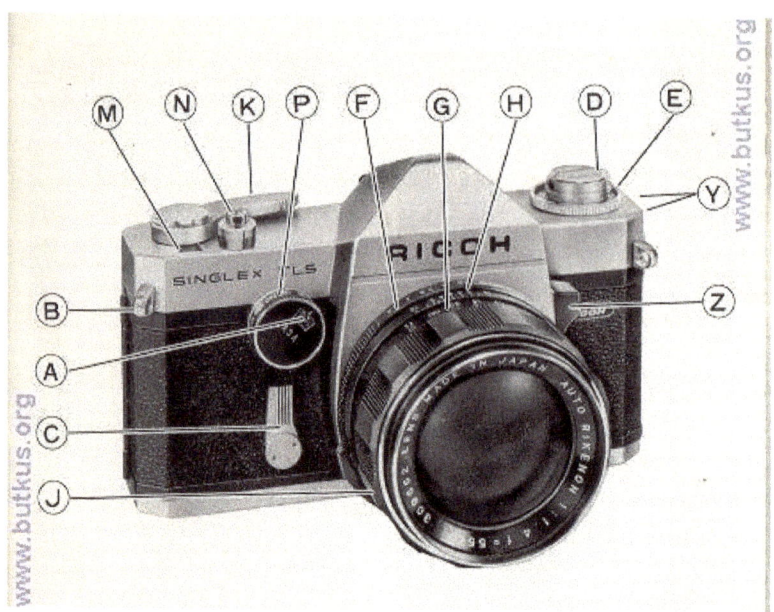

Figure 55 -   Billy Meier's Ricoh 55 mm camera. (Butkus.org).

Another point to note is that several people, among them Wendelle Stevens in his *UFO Contact from the Pleiades* books, have stated Meier's camera was stuck on "near infinity," or "one f stop from infinity," which is incorrect. There are no stops on the Olympus 35 ECR; it has a smooth sliding distance scale. The writers may have made such mistaken comments from casually looking at the photographs or from assuming Meier used his Ricoh camera that does have f stops. Why would they assume that? First, it is possible to use a 55 mm lens to take a shot like this, and it was the better camera of the two. In recent emails from Frehner in 2019, however, we learned that Meier said he never did use the Ricoh, except perhaps initially in around 1980 in an attempt to take a few pictures. The Ricoh was too tricky for Meier to handle because it required two hands to hold and operate it.

Significantly, the ECR, however, had a larger DoF than many popular cameras at the time, and this may have led to photo observers assuming that with such a depth of field the camera was not on infinity focus, but near infinity. Also, using a 55 mm lens on near infinity focus things start to come into focus at about 6 ~ 7 m on f16, and the DoF continues to infinity on f8 ~ f16 when focused between 8 ~ 15 m. The photo could show this, so such a conclusion is understandable. When enquiring through Christian

Frehner which camera he used, Meier insisted, however, that despite now having the Ricoh 55 mm camera he did use his old 35 ECR 42 mm to take this photo. Can it be shown that he did?

Focused on infinity, the ECR can focus from about 4 m away on f13, from about 5.24 m away on f11, and about 7.4 m away on f8 (see Table 2 below). This great DoF made infinity a standard manual default setting for many 35 ECR outdoor users, and the camera automatically decided the f stop setting. Set on infinity meant no focusing on the subject in this photo, which is what the photo shows. Furthermore, the rangefinder was not so accurate or quick and easy to use. Many users complained about it. According to Stevens in his *UFO Contact from the Pleiades: A Preliminary Investigation Report* (1982), Meier's camera had a "jammed focus and damaged viewfinder" (page 63).

The helical focus ring, or black distance scale on the lens barrel, is what is stuck on infinity with Meier's camera. (See Figure 54.) The "black" focus line on the camera's fixed front chrome ring indicates the centre index or distance at which the camera is in focus. On Meier's camera, it points directly at infinity. This thin black line in Figure 54 is, in fact, red. Perhaps the prior owner of the camera always used the camera on infinity, as many did, and it eventually jammed on it. It is not surprising then that Meier's 35 ECR was stuck there. (See also Frehner's video explanation and demonstration of Meier's 35 ECR Olympus on mycloud.ch.)

Stevens may also have assumed that early Meier researcher Ms Zinsstag of Basle, a relative of C. G. Jung, was correct when she said the camera was "jammed just a micro-adjustment short of the infinity position" (page 17). Presumably, that was an assumption on her part. Even if it were just less than a micro-adjustment off exact infinity, however, the camera would have almost the same DoF as a setting on infinity. Zinsstag, like Stevens, and as all human beings do, made an error. She says, for example, that the camera was an Olympus CR, but it is an ECR. We know this is an error because she correctly gave the ECR's serial number as 200519 (page 27) which agrees with Stevens (page 400). The number can also be made out upside down, and cropped, on the top of Meier's camera in the zoomed photo below (Figure 56). The "9" is behind the shutter release button; it is also evident in Frehner's online video demonstration of the camera. Perhaps Ms Zinsstag assumed the camera was a "micro-adjustment" short of

They are Here

infinity because she was told so by someone, or the previous owner assumed so, or because the centre index line on the dial is centred on the infinity mark and not its far end. The Olympus 35 ECR camera focus ring, however, does not move to the far end of the infinity mark on these cameras. Its extreme point on the dial, where it focuses on infinity, is when precisely centred on the infinity mark. The ECR setting, therefore, is on the infinity mark, just as in Figure 54 and the six-minute Frehner video.

So, given the evidence, it must be concluded that the Olympus 35 ECR jammed, focused on infinity. It is important to note that in photo #829 the distant horizon, the trees there, and the sky are what is in sharpest focus, precisely as they would be on infinity focus. This point, unfortunately, overlooked by most researchers to date is critical to revealing specific facts about the photo.

Figure 56 - Meier's Olympus 35 ECR serial number 200519 ("9" behind the shutter release button). Cropped Photoshop double "contrast" for clarity.

As noted earlier, the WCUFO width in the full-frame photo #829 is approximately 61% of the photo width, with about the front third or half of the craft out of focus, and the distant horizon in sharpest focus due to the camera's focus ring jammed on infinity. These are salient features to keep in mind. Also, it must be remembered that the 35 ECR owner's manual confirms that the camera only operates on f2.8 ~ f13 (and possibly a maximum of f13.5) and generally defaults to f8 or f11.

Which camera Meier used and the WCUFO size

First, looked at are the necessary camera distances from the centre of the craft for it to occupy about 61% of this full-frame picture width. Approximate possible distance ranges for the ECR and Ricoh cameras are estimated by looking at a 3.5-m-sized object (a surrogate WCUFO) and a 550-mm-sized object (the popular sceptic model size) through a) the Olympus 35 ECR viewfinder, and b) an SLR 55 mm lens on a 35 mm film camera. The distance ranges are derived by visually estimating 60 ~ 61% of the frame width and then measuring the distance to the object in question a few times to cover a few 60% estimations.

Since distance ranges double when doubling the WCUFO size, the 7 m distances are double the 3.5 m distances, and the 550 mm model distances are 15.7% (0.157) of the 3.5 m distances. We estimate this ad hoc estimation is subject to an eye judgement and ECR viewfinder inaccuracy of ±10 ~ 15%. Table 1 shows the results with approximate necessary camera distances from the centre of the crafts for them to occupy 60 ~ 61% of the photo width.

CAMERA LENS	550 mm model	3.5 m WCUFO	7 m WCUFO
ECR 42 mm	0.78 ~ 0.95 m	5 ~ 6 m	10 ~ 12 m
RICOH 55 mm	0.95 ~ 1.41 m	6 ~ 9 m	12 ~ 18 m

Table 1 - Approximate necessary camera distances from the centre of the craft for it to occupy approximately 61% of the photo width.

They are Here

These figures we now check on appropriate Depth of Field (DoF) tables and photo details to see possible explanations for how the picture was taken and by what camera.

Depth of Field (DoF)

As mentioned earlier, the depth of field is the distance parameters at which objects are in focus. The DOFMaster DoF tables for 35 mm film readings presented below in Tables 2 and 3 are the DoF tables courtesy of dofmaster.com for the 42 mm and 55 mm lenses. These tables give the data for the Olympus 35 ECR 42 mm and the Ricoh 55 mm cameras to help follow the investigation. Here is how our three objects in Table 1 show up using the three tables to hone down the possibilities. It is important to remember that for Meier's Olympus 42 mm camera jammed on infinity focus, only the bottom line in Table 2 is essential, while the 55 mm Ricoh could be on any focus setting.

A small model

The owner's manual from Orphancameras confirms the Olympus 35 ECR only operates between f2.8 and f13, and even if it could use f16, the camera would unlikely autoselect that on this overcast day. To occupy 61% of the picture width, the model must be 0.78 ~ 0.95 m from the camera, so no more than about 0.95 m away (Table 1). It is essential to remember that the DoF must extend to infinity as this is what the photo shows.

Consulting Table 2, no calibration makes a 0.95 ~ infinity DoF possible for a small model. The closest is a DoF of 1.93 m ~ infinity on f16 which is a bit beyond the camera's range of about 4 m to infinity focus on f13. At 4.0 m, however, the model is much too far away and too small in the picture. Furthermore, within the close range of 0.78 ~ 0.95 m, the model is entirely out of focus, when the horizon is in focus.

Even without the black focus ring stuck on infinity focus, the nearest point of focus would be over 2 m, and that is still way too far away. Any small model would, therefore, be entirely out of focus, and not what the photo shows, which is, the closest part of the craft out of focus, but its middle, far side, and the horizon in focus. So this photo was not taken using the 35 ECR camera with a small model.

What about a 550 mm model taken with the Ricoh 55 mm lens? Could this be possible?

	35mm film				Focal Length: 42 mm												
Distance	f/2		f/2.8		f/4		f/5.6		f/8		f/11		f/16		f/22	
in meters	Near	Far	Near	Far	Near	Far	Near	Far	Near	Far	Near	Far	Near	Far	Near	Far
0.25	0.25	0.25	0.25	0.25	0.25	0.25	0.25	0.26	0.24	0.26	0.24	0.26	0.24	0.26	0.23	0.27
0.5	0.49	0.51	0.49	0.51	0.48	0.52	0.48	0.52	0.47	0.53	0.46	0.55	0.44	0.57	0.43	0.61
0.75	0.73	0.77	0.73	0.78	0.72	0.79	0.7	0.8	0.68	0.83	0.66	0.87	0.63	0.93	0.59	1.03
1	0.97	1.03	0.96	1.05	0.94	1.07	0.92	1.1	0.88	1.15	0.84	1.23	0.79	1.35	0.73	1.58
1.5	1.43	1.58	1.4	1.61	1.36	1.67	1.32	1.74	1.25	1.87	1.17	2.08	1.07	2.49	0.96	3.42
2	1.88	2.14	1.83	2.21	1.76	2.31	1.68	2.46	1.58	2.73	1.45	3.21	1.3	4.28	1.14	8.11
2.5	2.31	2.73	2.24	2.84	2.14	3	2.02	3.27	1.87	3.76	1.7	4.74	1.5	7.55	1.28	46.2
3	2.73	3.34	2.63	3.5	2.5	3.76	2.34	4.19	2.14	5.02	1.91	6.96	1.66	15.4	1.4	∞
3.5	3.13	3.97	3	4.2	2.83	4.58	2.63	5.24	2.38	6.61	2.1	10.5	1.8	59	1.5	∞
4	3.53	4.62	3.36	4.94	3.15	5.47	2.9	6.46	2.6	8.67	2.27	16.8	1.93	∞	1.59	∞
4.5	3.91	5.3	3.71	5.73	3.45	6.46	3.15	7.88	2.8	11.4	2.42	31.6	2.03	∞	1.66	∞
5	4.28	6.01	4.04	6.57	3.74	7.54	3.39	9.56	2.99	15.4	2.56	109	2.13	∞	1.72	∞
5.5	4.64	6.75	4.36	7.46	4.01	8.75	3.61	11.6	3.16	21.4	2.68	∞	2.21	∞	1.77	∞
6	4.99	7.52	4.66	8.41	4.27	10.1	3.81	14.1	3.31	31.7	2.8	∞	2.29	∞	1.82	∞
8	6.3	11	5.79	13	5.19	17.4	4.53	34.1	3.84	∞	3.16	∞	2.53	∞	1.97	∞
10	7.47	15.1	6.76	19.2	5.96	31	5.11	238	4.25	∞	3.43	∞	2.7	∞	2.07	∞
15	9.94	30.5	8.72	53	7.43	∞	6.15	∞	4.94	∞	3.87	∞	2.96	∞	2.22	∞
20	11.9	62	10.2	500	8.48	∞	6.85	∞	5.38	∞	4.13	∞	3.11	∞	2.3	∞
30	14.9	∞	12.3	∞	9.88	∞	7.73	∞	5.91	∞	4.44	∞	3.28	∞	2.39	∞
50	18.5	∞	14.7	∞	11.4	∞	8.61	∞	6.41	∞	4.71	∞	3.43	∞	2.47	∞
∞	29.4	∞	20.8	∞	14.7	∞	10.4	∞	7.39	∞	5.24	∞	3.72	∞	2.64	∞

Table 2 - DoF (Depth of Field) for a 42 mm lens with 35 mm film. Details from DOFMaster 2005 ~ 2019. Dan Fleming. All rights reserved.

Again, to occupy 61% of the picture width, a 550 mm WCUFO model must centre at 0.95 ~ 1.41 m from the camera (Table 1). Still, when using a 55 mm lens, it is impossible to keep the back of the craft in focus because that would need a DoF ranging from around 1.0 m ~ infinity. Referring to Table 3, such a DoF is impossible for a 55 mm lens. The nearest possible DoF is 3.54 m ~ infinity on f16, the smallest aperture likely possible on this overcast day, but at this distance, the model again is far too small in the picture frame and still way too far away. Even on f22 the DoF is 2.37 m ~ infinity putting the 550 mm model entirely out of

They are Here

focus. Hence, the WCUFO was not a small 550 mm WCUFO model photographed by Meier with a 55 mm lens.

Distance in meters	35mm film Focal Length: 55 mm													
	f/2		f/2.8		f/4		f/5.6		f/8		f/11		f/16		f/22	
	Near	Far	Near	Far	Near	Far	Near	Far	Near	Far	Near	Far	Near	Far	Near	Far
0.25	0.25	0.25	0.25	0.25	0.25	0.25	0.25	0.25	0.25	0.25	0.24	0.26	0.24	0.26	0.24	0.26
0.5	0.5	0.5	0.49	0.51	0.49	0.51	0.49	0.51	0.48	0.52	0.48	0.53	0.47	0.54	0.45	0.56
0.75	0.74	0.76	0.74	0.76	0.73	0.77	0.72	0.78	0.71	0.79	0.7	0.81	0.68	0.84	0.65	0.89
1	0.98	1.02	0.97	1.03	0.96	1.04	0.95	1.06	0.93	1.08	0.9	1.12	0.87	1.18	0.83	1.27
1.5	1.46	1.54	1.44	1.56	1.42	1.59	1.39	1.63	1.35	1.69	1.29	1.79	1.22	1.95	1.13	2.22
2	1.93	2.08	1.9	2.12	1.86	2.17	1.8	2.24	1.73	2.36	1.64	2.56	1.53	2.89	1.39	3.55
2.5	2.38	2.63	2.34	2.68	2.28	2.77	2.2	2.9	2.09	3.1	1.96	3.45	1.8	4.08	1.61	5.54
3	2.83	3.19	2.77	3.27	2.69	3.4	2.57	3.59	2.43	3.91	2.25	4.48	2.04	5.63	1.81	8.85
3.5	3.28	3.76	3.19	3.87	3.08	4.05	2.93	4.34	2.75	4.82	2.52	5.71	2.26	7.72	1.97	15.4
4	3.71	4.34	3.6	4.5	3.46	4.74	3.28	5.14	3.05	5.82	2.77	7.18	2.46	10.7	2.12	34.9
4.5	4.14	4.94	4	5.14	3.83	5.46	3.6	5.99	3.33	6.95	3	8.98	2.64	15.3	2.25	1783
5	4.55	5.54	4.39	5.81	4.18	6.22	3.91	6.92	3.59	8.23	3.22	11.2	2.8	23.2	2.37	∞
5.5	4.96	6.17	4.77	6.49	4.52	7.02	4.21	7.92	3.84	9.68	3.41	14.1	2.95	40.4	2.48	∞
6	5.37	6.8	5.14	7.2	4.86	7.85	4.5	9	4.08	11.4	3.6	18	3.09	106	2.57	∞
8	6.91	9.5	6.54	10.3	6.08	11.7	5.53	14.4	4.91	21.6	4.23	74	3.54	∞	2.87	∞
10	8.35	12.5	7.82	13.9	7.17	16.5	6.42	22.6	5.59	47.4	4.73	∞	3.88	∞	3.09	∞
15	11.6	21.3	10.6	25.8	9.42	36.8	8.16	93	6.86	∞	5.6	∞	4.45	∞	3.45	∞
20	14.3	33.1	12.8	45.4	11.2	96	9.44	∞	7.74	∞	6.18	∞	4.8	∞	3.65	∞
30	18.8	74	16.3	187	13.7	∞	11.2	∞	8.89	∞	6.88	∞	5.22	∞	3.89	∞
50	25.1	5345	20.8	∞	16.8	∞	13.2	∞	10.1	∞	7.57	∞	5.6	∞	4.1	∞
∞	50	∞	35.7	∞	25.3	∞	17.9	∞	12.7	∞	8.97	∞	6.36	∞	4.51	∞

Table 3 - DoF (Depth of Field) for a 55 mm lens with 35 mm film. Details from DOFMaster 2005 ~ 2019. Dan Fleming. All rights reserved.

These inescapable facts dictated by the optics of the camera tell us that a small model WCUFO shot with either the Olympus 35 ECR camera or the Ricoh 55 mm camera is not what photo #829 shows.

Recent tests (2020) by Christian Frehner, taking photos at this site with an Olympus 35 ECR camera, show, by comparing the field of view on the photos, that Meier probably used the Olympus camera here, not the Ricoh. They also show that, with the distant landscape in very clear focus as it is, a nearby model would be out of focus because, again, Meier's Olympus camera was stuck on infinity focus.

Below the WCUFO with Trailer and Generator

For a 3.5-metre WCUFO

Note: To occupy 61% of the picture width (Table 1), the 3.5-metre craft shot with the ECR 42 mm lens needs to be about 5 ~ 6 m away and shot with the Ricoh 55 mm 6 ~ 9 m away. For camera data, refer again to DoF Tables 2 and 3 above from DOFMaster. While we only need the bottom line for Meier's Olympus camera which is always focused on infinity, the tables include from 0.25 m to illustrate the extremely short DoF there, underscoring the impossibility of the object photographed being a small model:

With a 42 mm lens (Table 2): Again, focused on infinity, this time referring to Table 1, the DoF must be around 5 ~ 6 m extending to infinity. Table 2 shows this is possible on f8 ~ f11. So the object could be a 3.5-metre WCUFO centred at around 5 ~ 6 m away. The front of the WCUFO is then about 3.3 ~ 4.3 m away.

With a 55 mm lens (Table 3): The DoF needs to be around 6 ~ 9 m to infinity, for a 3.5-metre WCUFO. For the Ricoh, not stuck on infinity focus, on this overcast day, it is possible at f8 ~ f11 at 6.9 m to 8.9 m away (Table 3). So the object could be a 3.5-metre WCUFO centred at around 7 ~ 9 m away making the front of the WCUFO about 5.15 ~ 7.3 m away from the camera.

So the craft in photo #829 could be a 3.5-metre diameter WCUFO centred at 5 ~ 6 m away shot with the ECR 42 mm lens, or centred at 7 ~ 9 m away for the 55 mm Ricoh lens. So on this somewhat overcast day, *either camera could have taken the photo.*

For a 7-metre WCUFO

To occupy 61% of the picture width, shot with the ECR 42 mm lens the 7-metre craft needs to be 10 ~ 12 m away and with the 55 mm Ricoh lens, 12 ~ 18 m away (Table 1). Camera calculations are similar to those for the 3.5-metre WCUFO.

With a 42 mm lens (Table 2):

So, a 7-metre craft needs a DoF of 10 ~ 12 m extending to infinity. A setting of f4 ~ f5.6 is possible for a 7-metre craft since the DoF is around 10 ~ 14 m to infinity with the camera focused on infinity and the craft centred at approximately 10 ~ 12 m away. So the front of the WCUFO is about 6.5 m ~ 8.5 m away.

They are Here

<u>With a 55 mm lens (Table 3)</u>: The DoF needs to start at around 12 ~ 18 m and extend to infinity for a 7-metre craft. Table 2 shows an aperture setting of f4 ~ f8 is possible for a 7-metre craft since the DoF is around 12 ~ 18 m to infinity with the craft centred at about 12 ~ 18 m away. Its front is then between 8.5 m and 14.5 m away from the camera.

So the craft in photo #829 could, in theory, be either a 3.5-metre diameter or a 7-metre diameter WCUFO shot with either a 42 mm lens or a 55 mm lens. Either camera and either craft are reasonable.

However, the *ECR 42 mm lens shot of a 3.5-metre WCUFO at 5 ~ 6 metres away from its centre fits exactly Meier's claim that he shot the WCUFO at about 5 ~ 6 m away*, and he probably shot it automatically on f8.

For a 14-metre WCUFO

Observing the ground, we note a minimum DoF range from about 3 m in front of the trailer, or about 22.5 m to infinity.

A 14-metre WCUFO extrapolated from Table 1 must be centred at 20 ~ 24 m away for a 42 mm lens and 24 ~ 36 m away for a 55 mm lens. We now look into these two possibilities.

<u>A 42 mm lens</u>: The camera formula shows momentarily that the trailer is about 19.5 m from a camera with a 42 mm lens. The generator and narrow road path in front of the trailer are in focus and account for about 3 m. So the DoF must be at least 17 m to infinity when using the Olympus 42 mm camera. Referring to table 2, the only aperture that gives a DoF of 17 ~ 20 m to infinity is f 2.8, which is an improbable automatic exposure choice for this picture. The Olympus 35 ECR usually shot automatically on f8 or f11 and while the trees in this shot are dark, the sky is light, and both occupy about one-third of the photo. In the original photo #829 (Figure 52) some clouds and the horizon are in focus, and the remainder of the picture is relatively dark, which tends to dismiss the possibility of an f2.8 aperture which would allow more light into the camera. Conversely, perhaps the dark scene led the camera to select the f2.8 stop to compensate?

It may make no difference which, since at 20 m away, the distance that camera formula calculations inform is the distance

Below the WCUFO with Trailer and Generator

to the trailer, a 14-metre WCUFO must tower over the entire trailer blocking all the bright overcast skylight from shining on its roof. Refer to Figure 57 and imagine a WCUFO twice the size of the largest one there centred at 20 m away. Such a vast WCUFO of 14 m right over the trailer, as it would be, should block all direct overhead skylight from shining directly onto the roof, but it does not. It could be, however, as Dyson Devine pointed out to us, that if only a narrow corridor of light is open to the camera, the WCUFO would be invisible to anyone else present and allow the overhead light to shine through the almost entirely invisible craft. We do know, and momentarily show, that the WCUFOs at night utilized a narrow corridor of light open only towards Meier's face and camera which would render the craft invisible to anyone else present. The big question is whether that is also happening here because it does seem to be the Plejaren's preferred modus operandi.

A look at Figure 48 and photo #844, however, shows the daytime WCUFO blocking the light from the sun onto the proximal tree. So this broad line of light must have been open as well as one to the camera, and it is an extensive area of light casting a wide shadow, suggesting perhaps that since no one was around the craft was allowed to be fully visible at the time. Maybe even for some reason, the WCUFO were always fully visible in daylight but not at night.

Curiously, Quetzal, the Plejaren in the craft, reportedly had to get into it and leave and repeatedly return due to the number of people that would appear at the site. If the craft were invisible, why would this be necessary?

So which was the case in our photo #829? Theoretically, it could be either.

There is also the matter of the tree fronds over the edge of the WCUFO (discussed momentarily) that look considerably more than the one or two meters in front of the trailer necessary for a craft centred at 20 m away. They place the craft closer to the camera. Taken together with the unlikely f2.8 stop for this photo, we conclude with reservation, this is not a picture taken of a 14-metre WCUFO with the 35 ECR Olympus 42 mm lens.

<u>A 55 mm lens</u>: Referring to Table 3, this is marginally possible on f2.8 ~ f4, but the WCUFO then is centred from almost precisely

over the front of the trailer and extending right over it, to being situated just beyond the trailer, which is 25.5 m away by camera formula calculations (see below). In photo #828, a comparison of the trees where the trailer's shadow casts towards them, with the fronds over the closer WCUFO edge enables an approximation of the distance to the trailer (Figure 59). These tree distances suggest the WCUFO is between the trailer and the camera.

Furthermore, with the trailer and 3 m of foreground before it in focus, the WCUFO front edge must be some metres closer than the trailer. At 14 m in diameter and 24 m away, however, the large craft would be entirely over the trailer again blocking any overhead light from reflecting off its roof if the craft is only visible in a corridor of light towards Meier's camera. Moreover, its front edge would be in clear focus, not blurred as in the photograph. The camera formula informs that the trailer is 25.5 m away for a 55 mm lens and given about 3 m of additional foreground in focus the DoF would begin at a minimum of 22.5 m to infinity. Hence the entire 14-metre WCUFO should be in focus.

The Ricoh 55 mm camera on f2.8 or f4 has a DoF of 20.8 m or 25.3 m respectively to infinity and centred at the minimum 24 m away more than half of the WCUFO is out of focus on f4, while less than half of it is out of focus in the picture. While the DoF works for f2.8 focused at 50 m, this aperture setting might, again, seem too open for this environment; however, it might be possible. F 5.6 focused on infinity gives a DoF from 18 m ~ infinity, which means only the first one metre of the WCUFO would be in focus, which seems too little, but again, perhaps it might work.

A problem with these two possibilities is that at the minimum distance of 24 m away, the craft still hovers entirely over the trailer at 25.5 m away. While this should block all direct overhead skylight from reaching its roof, it does not. Furthermore, the WCUFO height is not much more than that of the trailer roof, strongly suggesting that the WCUFO is not exactly over the trailer; otherwise, it would block all direct skylight from shining onto its roof. Again, however, this is irrelevant if the craft is only visible towards Meier and his camera. The trailer roof noticeably reflects the light of the overcast sky. The WCUFO also does not block light from the overhead sky reaching the generator in front of the trailer suggesting that the WCUFO rear is within 24 m of the camera or it

Below the WCUFO with Trailer and Generator

is mostly invisible or translucent. Again, which is it? Either explanation is possible.

More significantly, however, the camera formula puts a 14-metre WCUFO at 35.7 m (Table 4) where the craft will be both entirely beyond the trailer and entirely in focus, which is not what the photograph shows. So the camera formula tells us that this is not a 14 m WCUFO shot with a 55 mm lens.

So we conclude this is also not a 14-metre WCUFO taken with Meier's Ricoh 55 mm lens camera.

Through the Rangefinder

The rangefinder photo parameter readings are not entirely accurate and give *approximate* ideas of the photograph parameters. We, therefore, only use it to estimate 60% of the picture width, which we estimate has a ±10 ~ 15% error factor.

Conclusion

First, as initially established, the object photographed cannot be a small model. The photograph informs us that whichever lens Meier used, focusing anywhere up to 1.4 m from the camera creates a hopelessly short DoF rendering everything from 2.5 m away and up to the horizon very much out of focus. Of course, that is not what the photograph shows; it shows a large DoF with the horizon the point of focus, confirming a lens focused on infinity. Also, focused on infinity, the closest possible point of focus is over 3.7 m away on f16, at which distance a small model becomes far too small in the picture.

Second, the object could be a 3.5-metre diameter WCUFO centred at 5 ~ 6 m away shot with Meier's Olympus 42 mm ECR or centred at 7 ~ 9 m away shot with his Ricoh 55 mm lens. For the Olympus 42 mm lens, the f stop is around f8 ~ f11, and for the Ricoh 55 mm lens, the f stop is also around f8 ~ f11.

Third, the object could be a 7-metre diameter WCUFO centred at 10 ~ 12 m shot with Meier's Olympus 35 ECR or centred at 12 ~ 18 m away shot with his Ricoh 55 mm lens. For the Ricoh 55 mm lens, the f stop is around f4 ~ f8, and for the Olympus 42 mm lens, the f stop is around f4 ~ f5.6.

Finally, this is very unlikely to be a 14-metre WCUFO because the craft would be too far away, either over or behind the trailer and this would block the sky from being reflected on the trailer roof. Also, we notice fronds over the WCUFO edge (discussed momentarily) that put the WCUFO between the camera and the trailer. The WCUFO would also be so far away as to make it in complete focus which it is not in the photograph (Table 1).

Again, the ECR 42 mm lens shot of a 3.5-metre WCUFO centred 5 ~ 6 m away with its front at 3.25 m ~ 4.25 m away from the camera is the closest to Meier's claim that he shot the WCUFO at about 5 ~ 6 m away using his 42 mm Olympus 35 ECR camera. His claim fits the photo facts and data perfectly.

Camera formula confirmation

The camera formula is used next to calculate the WCUFO's distance from the camera and hopefully determine which possibility above is right, and especially, which WCUFO the photograph shows: the 3.5-metre or the 7-metre craft.

Photographers use the camera formula to calculate distances of objects in a photo once the real or actual size of another object in the photo is known. Here this is used to calculate the distance from the WCUFO and small models to the camera. The camera formula is $h/H = f/D$ or:

$$D = H (f/h) \text{ where:}$$

D = Distance to the object (Meier's trailer) from the camera

H = real Height of the trailer, or Width in our case (not subject to any perspective they are approximately equal, so either is useable).

f = the focal length of the lens

h = height/width of the trailer on the 35 mm film.

Based on the apparent 1.75 m width of the green trailer, and the WCUFO sizes, Table 4 shows Meier's distance from the WCUFO calculated from the camera formula given the exact 61.6%

ratio of the WCUFO width to the picture width. The table gives figures for both the 55 mm lens and 42 mm lens shooting a 550 mm model, a 3.5 m, a 7 m, and a 14-metre WCUFO.

Camera Formula distance calculations: example of the camera to trailer distance.

For a 55 mm lens:
$$D = 1.75 (55/3.77)$$
$$D = 1.75 \times 14.6$$
$$\mathbf{D = 25.55 \ m}$$

For a 42 mm lens:
$$D = 1.75 (42/3.77)$$
$$D = 1.75 \times 11.14$$
$$\mathbf{D = 19.5 \ m}$$

OBJECT	WIDTH on SCREEN (pc units)	WIDTH on FILM (mm)	REAL SIZE (m)	DISTANCE from CAMERA (f=55) (m)	DISTANCE from CAMERA (f=42) (m)
PHOTO	669	35.0			
TRAILER	72	3.8	1.75	25.55	19.5
0.55 m Model base centre	412	21.6	0.55	1.40	1.07
Near edge				1.13	0.79
Far edge				1.68	1.35
3.5 m WCUFO centre base	412	21.6	3.5	8.93	6.82
Near edge				7.18	5.07
Far edge				10.68	8.57
7 m WCUFO centre base	412	21.6	7.0	17.86	13.64
Near edge				14.36	10.14
Far edge				21.36	17.14
14 m WCUFO centre base	412	21.6	14.0	35.72	27.28
Near edge				28.72	20.28
Far edge				42.72	34.28

Table 4 -   Distances using the camera formula.

These calculations tell us that the trailer was either just about 25.55 m or 19.5 m away from the camera depending on the lens used (see Figure 57). From Tables 2 and 3, it is evident in theory that Meier could have used either the 42 mm lens or the 55 mm lens and those results agree with these from the camera formula. Specifically:

For a 3.5-metre diameter WCUFO shot with the 42 mm ECR lens, the centre of the craft would have been about 6.8 m from Meier's camera, putting its front at just about five metres from Meier. Using the 55 mm lens, the centre of the craft would have been at about 8.9 m from Meier and his camera. These findings are very close, differing by only 13% from the earlier DoF

Below the WCUFO with Trailer and Generator

calculations from Tables 2 and 3 of 5 ~ 6 m away and 6 ~ 9 m away respectively.

For a 7-metre diameter WCUFO shot with the 42 mm lens, the centre of the craft would have been about 13.6 m from his camera and using his 55 mm lens, about 17.9 m from his camera. These findings are, again, very close at only 13% from the earlier DoF calculations from Tables 2 and 3 of 10 ~ 12 m away and 12 ~ 18 m away respectively.

<u>If Meier used his Olympus 35 ECR 42 mm lens,</u> the trailer was about 19.5 m away, and *the front of a 3.5-metre WCUFO would have been about 5 m away from him, and the front of a 7-metre WCUFO about 10 m away from him* (plus or minus 0.5 m). Here the results differ by about 17% from the previous DoF based calculations of 3.25 ~ 4.25 m and 6.5 ~ 8.5 m respectively.

<u>If Meier used his Ricoh 55 mm lens,</u> the trailer would have been about 25.6 m away, *the front of a 3.5-metre WCUFO would have been about 7.2 m away* from him, and *the front of a 7-metre WCUFO about 14.4 m away*. Here the results concur with the DoF calculations of 6.5 ~ 8.5 m and 8.5 ~ 14.5 m respectively.

The camera method suggests that if Meier used his 35 ECR camera to shoot the picture at 5 m from the WCUFO as claimed, then it would have been the 3.5-metre craft he photographed. The front of a 7-metre WCUFO would have been 10 m away from him. However, according to DoF calculations, Meier could have shot a 7-metre WCUFO at 6.5 m away from its front edge, but from Table 1 its front edge needs to be at least 8.5 m away from him.

From the photograph alone, we cannot definitively establish, however, which camera Meier used or which large size WCUFO he photographed since both cameras give some possibilities as Figure 57 illustrates.

Considering the craft must occupy 61 ~ 62% of the picture width, Meier's claim of the 42 mm lens is the closest fit. If the front of a 3.5-metre craft was 5 m away from Meier, he shot the photo on f8 ~ f11. A 7-metre craft, on the other hand, according to the camera method had to be about 10 m away and shot on f4 ~ f5.6. Alternatively, using the DoF figures Meier is 6.5 m away from the front edge of a 7-metre WCUFO shot probably on f5.6 ~ f8.

While Meier's Olympus camera operates automatically from f2.8 ~ f13, it auto sets, so uncertainty exists as to the f-stop used.

Still, on this somewhat overcast day near the trees looking up towards the sky, he was probably on a camera stop of f4 ~ f11 and the camera usually auto-selected f8 ~ f11 (see Annex A, Olympus 35 ECR manual link).

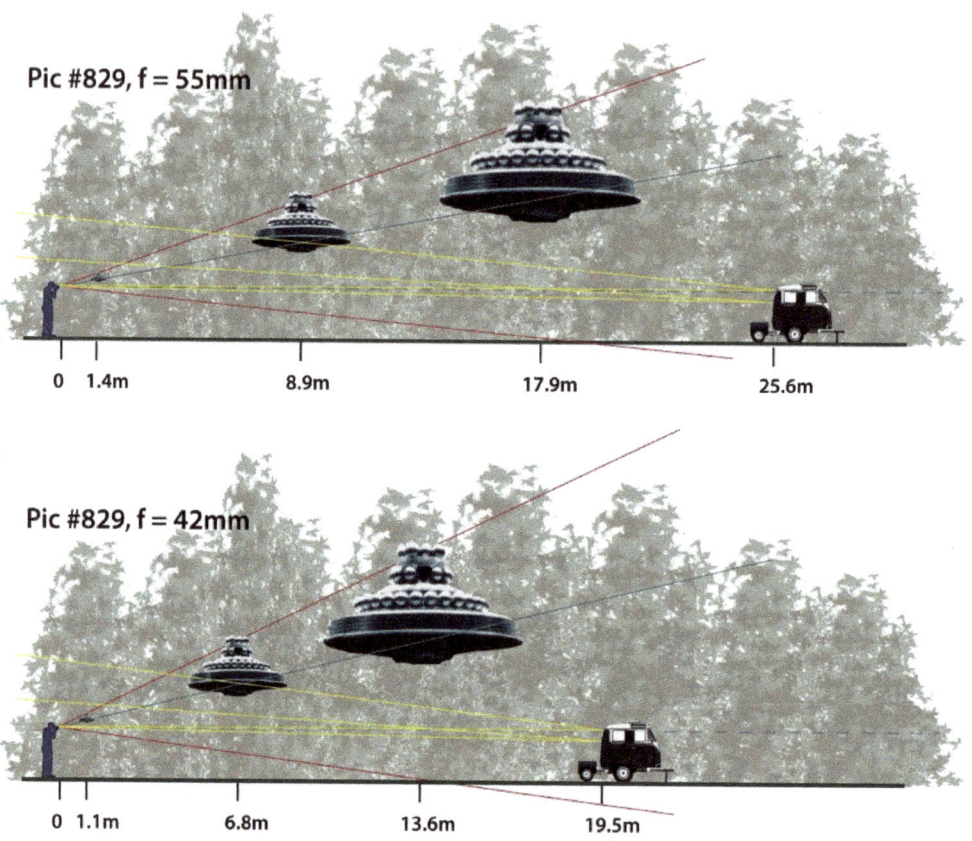

Figure 57 - WCUFO, trailer, and generator locations. Elevation views show possible positions for a 550-mm-sized model proximal to Meier, a 3.5-metre WCUFO, and a 7-metre WCUFO. Top: For the Ricoh 55 mm lens. Bottom: For the Olympus 35 ECR 42 mm lens.

An additional check

Now an alternative DoF check is performed using the Cambridgecolour online calculator. Entering the details, 35 mm full-frame, focal length 55 mm, f5.6, focus 200 m, gives a DoF of 15.57 m to infinity, again showing the possibility of producing this picture with Meier's 55 mm lens. It is very doubtful, however, that he would focus on 200 m while looking through the SLR camera lens at an object 5 ~ 15 m away. If he did, the WCUFO would then be about 10 m away from him.

For the 42 mm Olympus 35 ECR, the same calculator set on 35 mm full-frame, focal length 42 mm, f8, focused at 10,000 m produces a DoF of 6.89 m to infinity. It shows again that not only can Meier's 42 mm lens create a picture like this, but it fits well for a 7-metre diameter WCUFO. Also, since the DoF starts at 9.83 m away on f5.6 and that is too far away to have the back of the 3.5-metre WCUFO in focus the f stop automatically chosen was f8, which again fits well with this overcast sky and the camera's default mode.

The 3.5-metre WCUFO fits the data even better than the 7-metre one, and again, it agrees with Meier's claim that the craft was 5.0 ~ 6.0 metres away.

If Meier "remembers" incorrectly and he used a 55 mm Ricoh Singlex (Figure 55), model theorists' problems compound because the object then must be around three metres minimum in diameter to satisfy all the WCUFO with trailer photo elements (Figure 52). Deardorff previously pointed all this out on his website when writing at a time when people thought Meier used his 55 mm Ricoh Singlex camera to take this WCUFO photo.

In enquiries to Meier via Christian Frehner, Frehner informed us that Meier just confirmed, and reconfirmed, in late 2009 that the camera used to take the above photo *was* his Olympus 35 ECR with a 42 mm fixed lens stuck on infinity focus.

To conclude: all the photo details show this shot was of a 3.5-metre or 7-metre WCUFO almost certainly taken with Meier's Olympus 35 ECR 42 mm lens from about 5 ~ 6.5 m or 10 ~ 13 m away depending on which size WCUFO he shot. Everything fits well with what Meier claims for this photo. More importantly, it best explains all the photo details.

Finally, this cannot possibly be a small model. With either camera, due to their DoF, a small 550 mm model one metre away with its front out of focus and middle and back in focus, as the photo shows, results in the distant horizon out of focus when it should show the horizon in perfect focus. *So this photograph shows a large object just as Meier described; either a 3.5-metre-*

diameter WCUFO approximately 5 ~ 6 m away, or a 7-metre-diameter WCUFO about 10 ~ 13 m away from him.

Features of photo #829

Fronds in front of the craft

James Deardorff first discovered that by increasing only the brightness of this photo, tree fronds extend over the left side edge of this craft, confirming again that the picture shows a sizable craft and indeed not a small model. Below in Figure 58 is our minimal Photoshop enhanced reproductions of the photograph for comparison revealing these fronds over the edge of the craft.

One sceptic's attempt to recreate the fronds over the craft had the model against a non-needle tree, typical in England, with *tiny soft fir fronds*. These are nothing like the large-needled Norwegian Spruce or fir trees in Meier's photograph #829 (Figures 52 and 58). Visitors to this Swiss site have confirmed the rugged nature of these trees. Small fronds and a small model somewhat out of focus can give a slightly *similar* scaled image reading to big Norwegian fir fronds and a sizable craft, but the images are entirely different. Placing a small model next to the firs in Meier's Swiss photograph #829 would result in an altogether ridiculous effect. The big Swiss fronds would reveal the tiny size of the model.

Figure 58 - WCUFO with trailer detail showing tree fronds over the edge of the craft.

Below the WCUFO with Trailer and Generator

Perhaps it might be suggested Meier used a portable variety of a small fronded fir tree for his photo. Is this probable, likely, or even possible? The full image Figure 59 is the complete, "brightness"-only enhanced original image from Michael Horn's *www.theyfly.com*. The "brightness" only function minimises pixel and image distortion that other processes can produce. The only changes are image brightness, and squaring due to pixel nature of the digital image is hardly perceptible.

Frond colours

Noticeably, the dark grey-brown colours of the blurred fronds covering the WCUFO edge, the colour of others nearby and those less blurred near the trailer, show the same variations in hue and tone. Also, the spaces of the actual blurred frond areas compared to spaces around and within them are similar to other blurred fronds and spaces in the top of the picture. The fronds' blurring is due to their proximity to the camera outside its DoF, and possibly wind movement. They are the closest things to the camera except for the front of the WCUFO.

Important things to note about objects close to the camera are: First, the closer an object to the camera, the more prominent and vibrant its colour appears, and it takes up a more significant picture area than the same sized object at a distance. Second, and less commonly recognised, the colour and tone of close-up objects are more vibrant and, if somewhat dark, darker than similar dark objects at a distance.

This effect is due to light scatter and atmospheric attenuation, and it enables us to estimate how far away objects are within a photograph. If the tree fronds were just pot-sized pieces in front of a small model, then the actual local colour of the firs would be showing up because they would be closer than 1 m to the camera for a small model. They would not be reading as the same dark, somewhat washed-out colour or hue and tone due to light scatter as the fronds near the trailer 20 ~ 26 m away. The local colours of firs at one metre would be distinctly visible and a different colour to ones lapping over the edge of the WCUFO at approximately 5 ~ 13 m away.

They are Here

Nr.829: 26. März, 1981; Säckler, Dürstelen   ©FIGU

Figure 59 -   Photo 829 from www.theyfly.com "brightness" enhanced only.

The size of spaces between the blurred fronds compared with similar fronds near the van at 19 ~ 26 m away suggests these blurred fronds are in the region of 1/4 ~ 1/2 the distance to the trailer, or about 4.5 m ~ 13 m away, given that the camera formula calculation demonstrated the trailer is about 19 ~ 26 m away. It is difficult to be more precise with the distance of the fronds because determining their exact size or the exact size of the pine needles or twigs is impossible due to indistinct focus. However, we can see they are large fronds on large trees.

If this were a small model, creating the fronds over the edge to match this would require having to paint, or colour any smaller fronded models or pieces precisely the same dark brown colour to fit the colour scheme of the photograph. Also, it would be necessary to ensure their local green colour did not show up revealing their nature as a photo prop. Smaller fronded firs are also generally a brighter green, compounding this problem of colour rendition and presenting the further problem facing the faker of having to know what colour they reproduce in a developed photograph to match. Hollywood or a talented artist could do this, given the time and enough attempts. However, looking at the perfection in shape, colour, tone and size seen in Figures 58 and 59, we think Meier would need to be a genius in art and photography to reproduce these effects to this degree of perfection.

To ensure the same colour and tone would require a similar degree of overcast sky conditions that dictate the exposure, light, and colour renditions. Anyone but the most accomplished artists would find such difficulty in producing those matchings to such precision as to withstand detailed computer analysis that it would require more than a stroke of genius, and many site visits. Furthermore, with a milieu of people around Meier at his home at any given time to spot such conspicuous visible activity, this possibility is remote enough to be ruled out. Moreover, it is clear from looking at these details that the fronds fall over the edge of the craft.

Amazingly, some people, even pro photographers, are still in denial of this. Maybe they do not appreciate the artistic ability needed to fake this, or perhaps psychologically they need an "exit door." Centuries from now, however, people are going to look back and wonder how so many people could be so blind as not to see and recognise the obvious: a literal 21st-century "I see no ships."

Significantly, these fronds were never even mentioned by Meier or noticed for years until Deardorff first pointed them out after adjusting brightness levels on his computer where he posted his WCUFO research.[7] Would any faker go to the immense trouble of producing an astonishingly tricky effect that would go unnoticed for years, and probably forever given the unknowing standpoint of the pre pc digital era in 1981, and never mention their thoroughly convincing evidence? Why painstakingly produce an effect requiring artistic and photographic expertise bordering on genius, when it could never be seen, and create it with the very intention of it not being visible? Such a supposition makes no logical sense at all.

Reflection idea untenable

Some sceptics give a casual mention to the possibility of reflection in a vain attempt to explain away these fronds. A close

They are Here

inspection of the image and area quickly disproves this weak suggestion.

A convergence line projected down from the WCUFO edge cuts the undercarriage in the middle of its first flange (see Figure 60), and the same must naturally occur on the left-hand side. The left-hand side, however, shows the tree fronds come right over this left edge obscuring it from view; and this does not happen with a reflection. A reflection could not possibly obliterate the edge, and these fronds create a new *irregular outer edge* to the craft left of the blue line in Figure 61. The irregular outer edge alone proves this is not a reflection.

A lighter reflection also would not cover up the darker horizontal lines curving around the front vertical edge of the WCUFO. These indisputable facts prove beyond any doubt that what we see is not a reflection but something covering and hiding the edge of the craft. Compare Figures 58 or 59 with Figures 60 and 61.

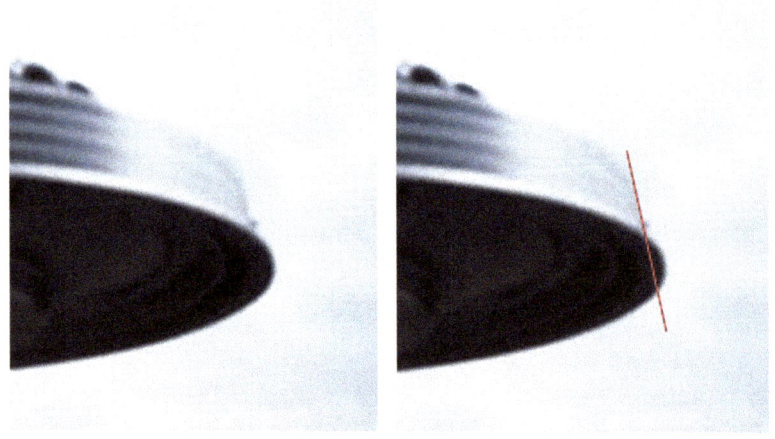

Figure 60 -   Convergence on the right side edge of the WCUFO.

Another weak claim has been that the fronds are a distortion due to pixelisation of the image. Pixelisation of this magnitude does not occur when merely using brightness or exposure on the image and there is no such other radical distortion seen anywhere in the picture.

Knowing that these massive fir fronds in Switzerland are lapping over the edge of this craft, we consequently know Meier's photograph is not of a 550 mm model but something considerably more substantial. The only question is how much more substantial; and it has been previously shown that it is either 3.5 m or 7 m in size. Of course, theoretically, it could be any size between 3 to 7 m in diameter.

In this photograph, the scale and colour of the many blurred fronds and how they compare with the more distant firs is consistent with the natural fir trees shown between 5 ~ 13 m from the camera. It is a safe conclusion that these are more of the same touching and hiding the edge of a large craft.

Figure 61 - Left-hand side of WCUFO showing the edge of the craft hidden by tree fronds.

In Meier's original photograph the object, occupying 62% of the photograph width, is out of focus at the front, in focus at its rear, and the photo has a DoF or range of focus from the centre of the WCUFO to the distant horizon. The problem this photograph #829 presents for the sceptic is that it is photographically impossible to produce all these photo features with a small 550 mm model.

Due to the laws of science and photographic optics, the object in this photo cannot be such a small 550 mm model. It is easy to prove to oneself how Meier used either an Olympus 35 ECR 42 mm lens or a 55 mm lens Ricoh SLR camera to shoot this photo (Figures 54, 55 and 62). Go out with a 42 mm and a 55 mm lens, obtain a circular surrogate object, and on an overcast day have:

a) the object occupying 62% of the photo width

b) the front half of the object out of focus

c) the back of the same object in focus and

d) the rest of the scene to the *distant horizon* in focus

e) the camera focus set on infinity or the far horizon

Any accurate photo recreation must satisfy conditions a) to e) that photo #829 evinces. Trying this, we discover it is impossible to recreate these conditions with any object less than 1.0 m ~ 1.5 m in diameter, at the very least. The DoF determined by the lens, or the field of focus within the photograph, does not allow the object to be any smaller than 1.0 m ~ 1.5 m. So it is a scientific fact that the craft under consideration must be at the very least 1.0 m ~ 1.5 m in diameter.

Finally, it should be mentioned that sceptics have made a good model approximation of the WCUFO of about 16-inches (406 mm), minus important specific details. Such a model, of course, however, does not mean Meier also made and used a 16-inch model – even if it was possible to use a 16-inch model to produce this photo, which, as shown from a technical standpoint, it is not. We have proven that even a 550 mm size model is hopelessly too small to reproduce the salient points of the photograph and to reproduce what this photo shows.

Figure 62 -   Meier with his Olympus 35 ECR camera in 1979.

WCUFO at Night Analysis

Introduction and key findings

While the pictures of the WCUFO hovering above Meier's courtyard and those above or close to the trees are indeed surprising, the WCUFO night photos show even more stunning and remarkable details.

- In photo #873, we noticed that if this WCUFO's proportions are the same as the others, then it can extend its central core upwards ¼ of a sphere diameter, meaning this WCUFO expands in height unless there are several WCUFOs with significantly different proportions.

- The nighttime WCUFO, shining orange in photo #873, exhibits a pink halo around it. Is this evidence of an ionised atmospheric plasma field? Attempts to explain the WCUFO as a small UFO model face several insolvable incongruences. Once more, the UFO model hypothesis fails.

- What look like distant street light reflections on the spheres do not behave like street light reflections. It seems Meier's claim is correct, that this WCUFO at night is *radiating* light rather than reflecting it from external light sources.

- In almost all these nighttime photos, the WCUFO is golden orange in colour or suffused by a golden orange glow. We investigate the possibility of this glow being a low energy plasma field and find that it most likely is.

- Another remarkable detail that neither the UFO scale model theory nor a conventional UFO explains is that only Meier's camera sees the WCUFO. This fact leads to the only possible explanation that the WCUFO is capable of hiding its light from illuminating its surroundings and via this screening allows only Meier and his camera to observe it through a narrow open corridor of light.

They are Here

- In some of the night pictures, the spheres seem to present a non-spherical form. We found this to be an optical effect due to the bright reflections between contiguous spheres, not an actual sphere deformation.

The extraordinary photo #873

Photo #873 is one of the most surprising and stunning pictures from the Billy Meier collection. At first look, it seems like a typical WCUFO photo. Some researchers, with a good sense of proportions, noticed its cupola seemed extended a small fraction upwards. Our investigations demonstrated that this is so (Zahi and Lock *Researching a Real UFO*; Zahi *Analysis of the Wedding Cake UFO* 74-page report).

Figure 63 - Photo #873. A WCUFO at night.

WCUFO at Night Analysis

The surroundings of this photo are very dark, so any external light that might illuminate this WCUFO is not illuminating anything else around, which makes exterior lights very unlikely.

The night of 28 July 1981, Meier was either standing behind his home or on another beamship as he photographed this 7-metre WCUFO slightly below him. Initially, reports were that he was flying in another WCUFO somewhere in central Switzerland. According to a March 2020 email from Christian Frehner, however, he was at the SSSC (Semjase Silver Star Centre), on his home property immediately behind his house. (See Figure 67.)

Curiously, Meier says this WCUFO radiates light. The "reflections on it", he says, are not from street lights but are instead lights radiating out from the WCUFO itself.

This photo was seen and known merely as in Figure 63, for more than 30 years. Recently, however, by using readily available technology for image processing, tools like Photoshop, new details are coming to light. Enhancing the brightness of this photo a few times reveals hidden details. Figure 64 shows some of them.

Figure 64 - Photo #873. An enhanced image of the WCUFO at night showing a wine-red ionised plasma sheath and a terrain below the craft.

Suddenly, a terrain is visible below the WCUFO with a grassy field sloping downhill. The path just outside and below Meier's house appears in the bottom right corner, and a wooden pole stands close to, maybe beneath, the WCUFO. No supporting devices exist to indicate it might be a UFO scale model. The wooden pole projects a unique shadow going downhill (see pole detail in Figure 65).

In colour, this photo reveals an orange or golden coloured WCUFO with a pink or wine-red halo around it. While we now discuss the red glow or probable "sheath of plasma" we return to discuss the golden orange glow a little later. Some sceptics attempted to explain this as a red curtain behind a UFO model. Momentarily, we discuss and find several reasons why that UFO model suggestion is not viable; but first, is this wine-red halo a plasma sheath of ionised air molecules?

A wine-red plasma sheath of ionised air molecules?

According to explanations of plasma sheaths by well-informed researchers and top scientists, the wine-red WCUFO "halo" or "plume" appears to have all the hallmarks of one. In his book *Unconventional Flying Objects: A Scientific Analysis*, well-respected ex NASA scientist Paul Hill explains the basics of this often observed phenomenon around UFOs:

> There is really no secret as to what this illuminated and illuminating sheath of atmosphere around the UFO is. It is a sheath of ionized and excited air molecules often called a plasma. It has all the many characteristics of ionized and excited air molecules and has no characteristics not attributable to ionized and excited air molecules with expected contaminants. Thus, the illumination is tied to an air plasma. I am not suggesting anything original as it has been suggested by many that such is the case. Indeed, any physicist who has made a study of UFOs must know they are characteristically surrounded by an air plasma (1995, pages. 53 ~ 54).

Apollo 14 astronaut, Edgar Mitchell, Sc.D., endorsed the top credentialed Hill and his book writing on its back cover:

> Paul Hill has done a masterful job ferreting out the basic science and technology behind the elusive UFO characteristics.... Perhaps this book will help bring solid consideration for making all that is known about extraterrestrial craft publicly available.

The pioneering scholarly work of renowned scientists like Hill is sadly unknown to those people today who have lost the appetite for good books and feed on nutritionless memes poured out daily into the cyberspace trough. It takes detailed reading, study, and research to arrive at the truth behind controversial matters like UFOs.

In the chapter of his book on "Illumination" Hill discusses the characteristic UFO plasma phenomenon in some depth explaining that when the plasma is energetic, it creates haze obscuring a clear view of the UFO. When the plasma emission is weak, however, the craft can be seen clearly (page 67). The WCUFO in photo #873 and all the night photos is without any obscuring haze. So if this is plasma, it has to be a low energy type. Is it? Hill notes that plasma colour ranges throughout the colour spectrum, but interestingly, high energy plasma is near the blue, and blue-white end and the low energy plasma colours are red and orange. Significantly, low energy plasma is most associated with low activity, low speed, and hovering or being static. This description fits the WCUFO. The WCUFO in these photos is indeed travelling slowly or even stationary, with low energy activity, so its plasma should indeed be red or orange; the photo #873 WCUFO has both.

We can probably conclude then, that the wine-red plasma sheath or plume is the lower energy form of the two. Hence, it probably relates to the main engine or power source of the craft. In contrast, the orange plasma probably relates to the smaller energy activators or "swinging wave accumulators" as the Plejaren call them according to Meier. This orange plasma may also relate to the many lights around and over the craft that we analyse momentarily.

The wine-red plasma plume above the WCUFO is probably either rising from the craft or plasma residue as the craft descended to where it is in the photo. To temporarily conclude

They are Here

commentary on this wine-red plasma sheath in photo #873 (we return to it briefly when considering why the WCUFO is a golden orange colour), researcher and Disclosure Project witness Dyson Devine observes:

> We can see this plasma halo really well on this cool, obviously calm, dark night. And the concomitant heat of the stationary, Plejaren, self-luminous craft makes a pretty plasma convection plume – if you can see what you're looking at – even without being photographed using schlieren equipment (Horn They Fly Blog. "What Are the Criteria...").

Figure 65 -   Photo #873. Detail of the wooden pole projecting a single shadow downhill below the WCUFO.

The size of this WCUFO

How big is this WCUFO? Is it possible to know? Yes, a reasonable estimate of its size is possible. Anybody with basic concepts of photography knows that night pictures like this one, are taken with the camera diaphragm wide open. Meier's automatic camera diaphragm must have been open when he took this photo, significantly shortening the depth of field. A short depth of field, in turn, means many objects at different distances go out of focus. Because the pole, proximal terrain, and WCUFO are all in focus at night-time, they share approximately the same location and distance from the camera, and the terrain detail evinces this.

Moreover, if the pole, which no longer exists at Meier's property, is the same distance from the camera as the WCUFO, their sizes relate. Typically, these fence poles are around one metre high. Furthermore, since, in this picture, the WCUFO diameter is seven times the height of the pole, we can conclude this WCUFO is about seven metres in diameter.

A 7-metre diameter WCUFO means there are at least two WCUFOs, one of 7 metres and the other 3.5 metres. This finding corroborates what Billy Meier claims, that he was photographing a big 7-metre WCUFO and that there are at least two sizes, and perhaps several sizes as Meier claims.

Our photo is a high-resolution, excellent electronic copy (TIFF file) taken from the original colour slide positive, which is why we see useful details. Is it possible to see something like this from pictures in old books? It is, but with a lower resolution. Figure 66 shows the same photo from lower-resolution photos available on the Internet and in old books by Wendelle Stevens.

This evidence has been lying there undiscovered for more than 30 years, in both low and high-resolution photos.

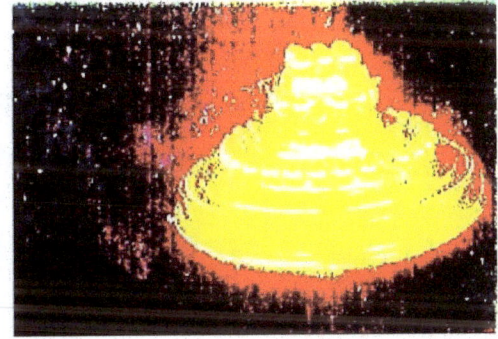

Figure 66 - Photo #873. Photo from old books. Left: The pole below the WCUFO is barely visible inside the circle. Right: The halo evident in old photos and a typical vertical pattern on the halo of low-resolution JPEG files.

They are Here

WCUFO location at the SSSC

At the back and towards one side of Billy Meier's home, is a small deck facing a terrain that extends downhill. Figure 67 shows two recent photos of this place.

Figure 67 - Photo #873 SSSC location. Top: View from the deck. Bottom: View uphill from the footpath towards Meier's home.

WCUFO at Night Analysis

Billy Meier might have been at the BM label position in Figure 67 (Bottom), to the left of the yard lamp looking towards the WCUFO label position, or looking down towards the photographer's location in the bottom photo of Figure 67 since there is minimal terrain shown in photo #873. Alternatively, Meier was standing on top of a companion beamship as claimed. Our research continues into the precise location of the WCUFO and Meier, but unfortunately, SSSC visits are postponed for everyone until the Covid-19 pandemic is over. Either way, he was looking downhill towards the field where today we see some small trees which were not there when photo #873 was taken.

A yard lamp affixed in a pole in this area labelled **1** and circled in the bottom picture is a source of light to Meier's left and accounts for the shadow projected from the pole extending downhill in this terrain. We can also see the footpath in these photos. At the time this WCUFO photo was taken, there was at least one wooden pole at the side of the path, and perhaps more elsewhere along its length.

UFO model suggestion untenable

Before this discovery, sceptics claimed this photo was of a UFO model sitting on a table, in darkness, which we now know is entirely wrong. There are no tables, or cables or anything suspending this flying object. More recently, some sceptics even suggested the pink halo to be a reflection of a red curtain behind a scaled UFO model. We investigate and finally thoroughly refute that suggestion below. The idea of a model WCUFO is untenable for the following reasons:

- The terrain below the WCUFO looks like a real natural field; there is nothing artificial about it. If fabricated, the terrain and pole must also be scale models. Both, the UFO model and the terrain model must, however, be at approximately the same distance from the camera; otherwise one of them would be totally out of focus. How could the UFO model hover in the air in 1981 a generation before drones became available? Even a flimsy thin wire would be visible in the enhanced photo which picks up remarkably hidden details. If the terrain were a model, the wire would be visible because any lamp or lamps used to illuminate the terrain must also light up any model strings or supports. Both the model and the terrain would be proximal to the camera at one metre away making the scale terrain model as visible as the UFO model, but that is not at all what the photo shows. In the small model proposition, due to their proximity, both the model and landscape should experience a similar degree of illumination, but they do not. Furthermore, why make a complicated model and not show or mention it for 40 years when it would bolster one's claim?

- If there were a red curtain behind the UFO model, the lamps would illuminate both the UFO model and the red curtain; but they do not. A black curtain instead? Then why is it wine-red? Why is the curtain visible just around the UFO model and not in other places? Why is the curtain not lit up by the lights? It is so close that it would have been. Why even hang a red foreground curtain when its exposure is more likely than any less likely to be exposed darker background? The suggestion of a curtain does not make any sense at all; furthermore, the "curtain" has significant gaps in it, preventing its alleged purpose. The sceptic red curtain evinces straw clutching to a pseudo-theory now sinking into oblivion.

- Bright reflections of at least two lamps are on ("in" according to Meier) the WCUFO spheres. Any model, therefore, requires two lights. However, there is only a single prominent pole shadow, not two.

- There are far too many problems with the UFO model idea for it to explain this photo. From a practical and scientific standpoint, Meier's explanation best fits the facts.

The big WCUFO hypothesis, however, also presents some challenges. If this is a big object, flying above an open field at the SSSC, why is the terrain below so dark considering all the WCUFO lights? Also, why is there only one single pole shadow, seemingly defying the laws of physics (which is impossible)? Meier explicates these challenges by discussing a curious screening device his UFOs have that is not unique to the WCUFO.

Meier claims the WCUFO *radiates* light. The suggestion is: brightness seen on the spheres is not from surrounding street lights, but a curious emission of light from this type of UFO. If the WCUFO radiates light, why is the terrain below so dark, since it is just two to three metres below? We also note that the craft's plasma sheath might illuminate the landscape. Also, if the WCUFO emission of light is not from a single point, such as a light bulb, but from the full WCUFO surface, it would not produce any shadow at all. So, why is there one single shadow from the pole going downhill?

In an old documentary by Japanese reporter Junich Yaoi (1985), we hear Meier's explanation. Yaoi asks Meier why or how only he was able to see and photograph several UFOs in action so clearly when they were close enough for many people to see. Meier explained the UFOs have a screening invisibility device that creates a surrounding field that hides them from curious people or onlookers. Here is how he described it in his broken English to Yaoi in his interview (now on YouTube):

Meier: *There is a free line only to see something through here...the camera. Or to my face, to my eyes* [gestures around his face and head, camera in hand]. *Then I'm staying here, get the picture...to get the picture from the ship. You stays [sic] there, by the lamp or by the tree, you can't see anything, because there, the sighting will be closed for you. The* [muffled over speaking but sounds like "doorway" or "hallway"] *will be open this way to the camera* [gestures slightly towards himself]. *This happens the most times* [sic].

Yaoi: *How could they do it?*

Meier: *I don't know.* [8]

So Meier could photograph them through this "hallway" of light while any other people around at the SSSC could not see them. So these UFOs could fly very near many people yet remain hidden from everyone's view except for Meier who says he was authorised to take photos of them.

With its invisibility device turned on, then, Meier would be able to see and photograph the WCUFO in photo #873, but no matter

They are Here

how close they were to it or Meier nobody else present would see it. In other words, *this WCUFO did not radiate any light outside of a narrow visual corridor toward Meier's camera*, and this would explain why the terrain below was so dark. No light from this WCUFO reached the ground. Any solitary pedestrian around at the time of night would not have seen anything. While this may sound like science fiction, it explains the otherwise inexplicable nature of this photograph: the WCUFO flying over an open field possessed and was, as Meier claims, using a screening device.

Finally, what about the single-pole shadow? We think the probable cause is the yard light shown in Figure 67 located on Meier's left that provides just enough faint light to illuminate the grass and footpath. Could Meier have perhaps used a flash, and his flashlight caused the shadow? Meier reported that he did not use a flashlight, and if he had, a bright blue dot would reflect visibly off the spheres. Also, the shadow's direction indicates the source was not very close to Meier's camera, perhaps 1.8 metres to his left, and this is the right distance we can see at the deck of his home. It is not clear how far the pole's shadow extends downhill. It ends out of visual range and so does not define the elevation of the light source. All we know is that because this shadow is very well defined, it cannot be caused by multiple WCUFO lights, assuming the UFO emits light. Of these possibilities, we think the most probable source of light for the shadow is Meier's yard lamp.

These are our current attempts to explain this photo showing a remarkable flying object seven metres in diameter with a screening device hiding it from the view of curious people that perchance might be around. It is radiating light and emits a flowing wine-red halo, which we think is almost certainly a plasma cloud from the WCUFO's primary energy source. We also note that the craft can extend its central core upwards, all while Meier rides onboard a companion WCUFO, or stands at his property's deck, photographing the craft from above. It may sound fantastic, but the photo demonstrates this fantastic nature. Sometimes the truth truly is stranger than fiction.

WCUFO vertical extension

The 74-page investigation report *Analysis of the Wedding Cake UFO* (Zahi) features a detailed analysis of this WCUFO central core extension. Two methods reveal a physically measurable vertical extension of 0.23 to 0.24 times the lower spheres' diameter (almost one quarter). Distance D-E on the following Figure 68 is half a sphere diameter, but comparing it to other WCUFO pictures, it should be only one-quarter of a sphere diameter.

So this WCUFO might have different dimensions or be capable of extending its central core upwards.

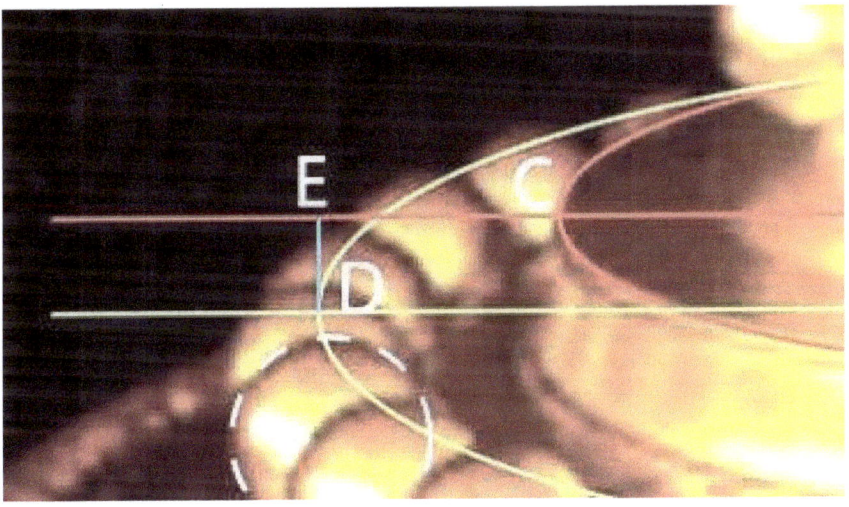

Figure 68 -   Upward extension of the WCUFO central core.

Does the WCUFO *emit* or *reflect* light?

Previously, we discussed that picture #873 is explained only by considering the WCUFO emitting light rather than reflecting it from distant street lights.

Initially, and naturally enough, a look at the nighttime WCUFO pictures would tend to suggest the lights on its spheres and cylindrical surfaces are the result of externally reflected lights. The sceptic tries to imagine a UFO model with cinematography lights placed around it. A non-sceptic supposes street lights surround a big WCUFO. Meier says the WCUFO radiated light and the "reflections" on the shining surfaces are not from street lights. He says he was there, saw this object moving in different locations, and was even flying with them. If indeed he was, his explanation must be right. In this section, the evidence is explored to verify in detail whether Meier's claim is correct.

Figure 69 -   Night-time sphere reflecting lights.

Figure 69 shows a reflecting sphere at night to test its reflections of surrounding lights. The first thing noticed is how tiny the reflected lights are. Comparing them with the "reflections" on the WCUFO, these tiny reflections of different colours lead to questions.

Why are the WCUFO lights so big, how big are they, and why are they the same golden orange colour? How can we locate these light sources, assuming they are external lights? There are no singular, or multiple "spotlights", like external electric street bulbs that can produce them. They evince a broad light source form, sometimes from several lights concentrated in a wide area.

A simple method exists to determine the direction of any reflected image on a sphere or cylindrical surface. The WCUFO has several spherical surfaces of different sizes, located on three levels; several cylindrical surfaces, like its broad base edge; and conical surfaces, like the upper platform with the little star-like forms engraved on its surface. Experiment 8 in our guidebook *Researching a Real UFO* describes in full a method to estimate the direction of any object reflected on a sphere, including, of course, the WCUFO spheres. Here, we briefly describe this method.

Figure 70 shows two templates for plotting reflected light on both spherical and cylindrical surfaces. Left, right, up, or down refer to the direction as seen from the WCUFO looking towards Meier's camera. Vertical curves (meridians) measure horizontal angles, and horizontal curves (parallels [9]) measure vertical angles on the spheres. On cylindrical surfaces, of course, it is only possible to measure horizontal angles.

Transposing the WCUFO sphere and cylindrical surface bright spots onto the templates reveals their horizontal and vertical angles as if looking at them from the WCUFO. These angles measure from a line connecting the WCUFO reflecting surface to the camera (baseline).

For example, looking at the bright spot marked **2** on the sphere in Figure 71, and transposing it onto the Sphere Template we find its location at 10 degrees elevation and about 30 degrees to the left. Meaning, from the sphere's location, we see a bright source of light, perhaps a street light, at an angle of 30 degrees left of Meier's camera and an angle of 10 degrees above Meier's camera.

They are Here

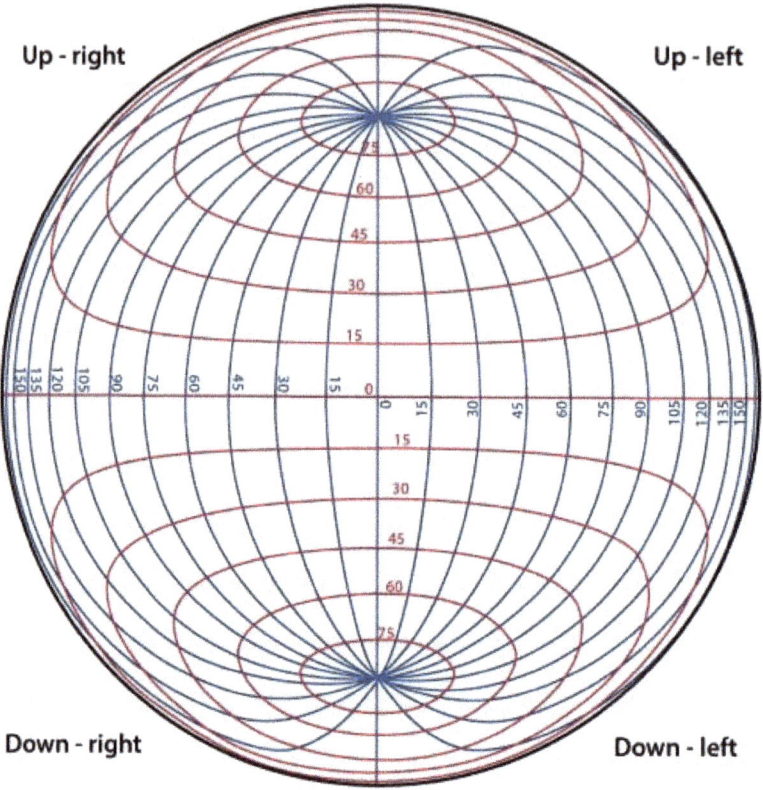

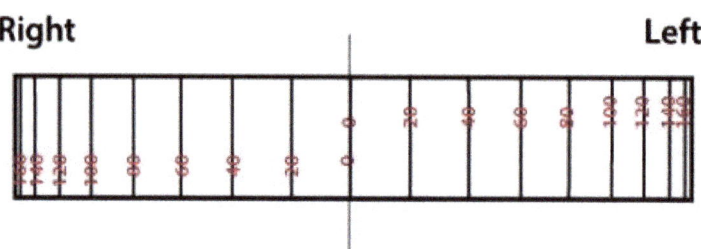

Figure 70 - Top: Sphere Template for plotting objects vertically and horizontally. Bottom: Cylinder template for plotting objects horizontally.

WCUFO at Night Analysis

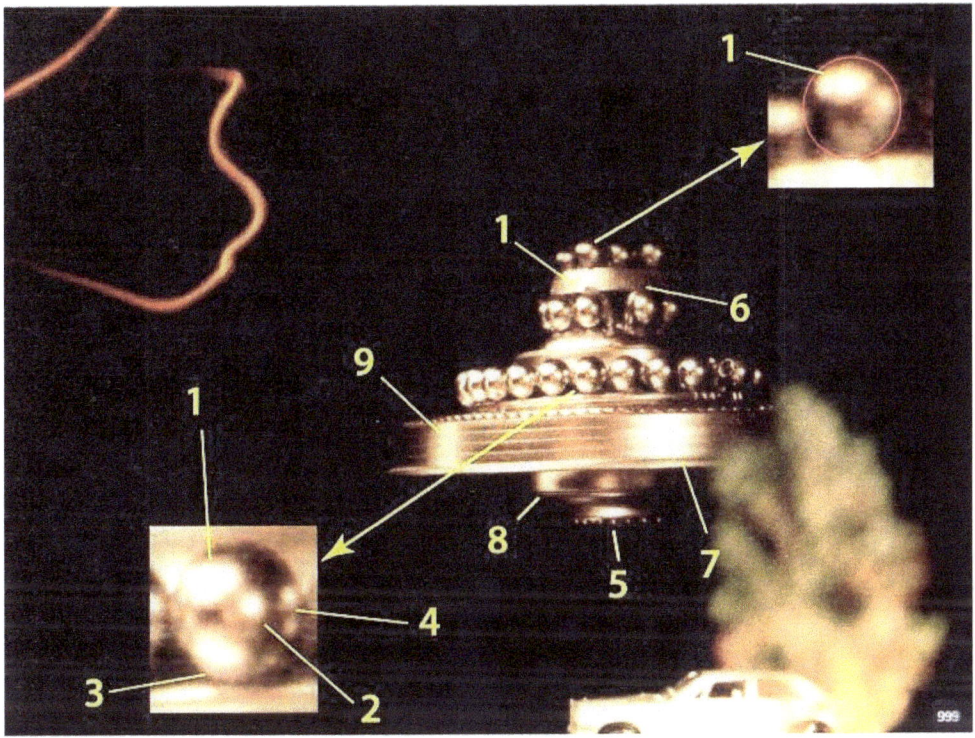

Figure 71 - Photo #999 from the book *Photo-Inventarium*.

Observing the WCUFO sphere and body reflections enable location estimates for any external light sources if there are any. If so, the reflections are congruent; they follow a familiar pattern. However, we notice several inexplicable incongruences in some of the photos that might suggest various explanations, from external lights of some kind, perhaps street lights, to big WCUFO lights, or maybe even studio lights of a UFO scale model.

Figure 71, photo #999, is from a series of WCUFO photos Meier took on the night of 2 August 1981. Meier says the WCUFO is behind the Mercedes Benz, and the out of focus tree is between the camera and the car. The WCUFO in the background is in good focus and remembering that Meier's camera is stuck on infinity focus, makes the somewhat distant WCUFO the clearest in focus. Here we notice, as mentioned before, that night pictures have a minimum depth of field producing images of objects out of focus at widely varying distances from the camera. The car in photo #999 is slightly out of focus, while the tree which is closest to the camera is wholly blurred.

They are Here

The bright curved line at the top left is caused by a telemetric disk, according to Meier. He says it is a little flying Plejaren monitoring device rapidly moving as he shoots the photo. So this photo shows a tree proximal to the camera entirely out of focus, a Mercedes Benz farther away but closer than the WCUFO which is seven metres in diameter when compared with the size of the car, and a bright trace of a telemetric disk.

What else is around? Analysing the WCUFO light "reflections" informs us.

Figure 71 shows zoomed images of two spheres, one at the top and one at the bottom. The numbers marked **1** through **9** indicate nine different "reflected" lights and their sources. "Reflection" **1** is of a huge bright light. It could consist of five horizontal lights in an area ranging from 10° to 120° right viewed from the WCUFO, and vertically from 20° to 80°. So this very bright light source, or group of five lights, covers an extensive area. The photo suggests their source would be from the top-left direction, above the telemetric disk. The lights in "reflection" **1** are from this source.

In the same Figure, **2** is a small bright spot that might well be caused by a distant street light. Its position is standard street light height and has an average pole height elevation of 10°, and it is 30° left as seen from the WCUFO. So, this electric bulb is to the right of Meier. The bright reflections on the vehicle are consistent with this source of light.

Number **3** is the reflection of light **1** on the WCUFO base, while **4** is the reflections of **1** and **3** from the nearest sphere, and **5** may come from the same source as **7**. However, it is very close to the WCUFO central axis and cannot be explained by a reflection from the vehicle since the far side of the automobile is probably dark. Perhaps it is coming from bulb **2**? Number **6** is a little mystery: it is very dark, and light from bulb **2** should be there. Perhaps, being conical, the surface's steeper receding vertical angle does not reflect the bulb. Also, the source is coming horizontally from the street light that may brightly illuminate **7**, but the light may be unable to reach up to **6**.

Numbers **8** and **9** on the WCUFO bottom and base are again a bit of a mystery. With the WCUFO covering a significant portion of the sky, if the light is from above, why is light or reflection **8** on the recessed WCUFO base? How can a light source from above

WCUFO at Night Analysis

illuminate this sheltered WCUFO base? The lowest of the five lights in **1** is 10° above the horizon, so it cannot illuminate this WCUFO base. Is a source of light coming horizontally from the WCUFO's left, or from below 100° to the right seen from the WCUFO? If so, why is it not visible in the top sphere reflections that should have a clear view of it?

Figure 72 -   Auto detail, photo #999.

A bigger question, given the WCUFO covers 25% of the sky, and all the lights "reflected" in the sphere's upper left, is, "Why is the tree so faintly illuminated?" The many lights should brightly illuminate it. The tree seems to be receiving light from the street light only, why not from the other sources?

Figure 72 shows a photo #999 detail of a parked Mercedes Benz. Notice the reflections from the street light, number **2** in Figure 71, are visible on the wheels and close to them. There is another little bright reflection on the windshield that may come from the WCUFO if its restricted corridor of viewing is not operating at this time. However, if there are big lights on the top left side, why do they not produce noticeable reflections on the front of the vehicle? They surely should, unless the source from these bright lights covering a significant portion of the sky on the top left side are not external light sources. Perhaps, as discussed earlier, this WCUFO is *emitting* light rather than *reflecting* it and all light from it is blocked except for the thin corridor of light to Meier and his camera. If so, the only reflection on the WCUFO from an external source might be the street light **2**.

Figure 73 shows a more prominent and brighter reflection of the light from the WCUFO on the car windshield. This figure is photo #1000 taken at 02:19 a.m. just one minute after #999. Meier has walked towards the car and the WCUFO. The top little green mark might be one of the tree leaves, making Meier perhaps

They are Here

just below the tree. The automobile is now closer, bigger, and becoming totally out of focus due to the depth of field effect with the camera set on infinity focus. The WCUFO is bigger than the car, which is closer, but the perspective makes them look the same size. Notice the "reflections" of the big lights on the top left side are now in a different place: They are now lower. The street light reflection is now in the same position as one of the big "reflections," making them indistinguishable.

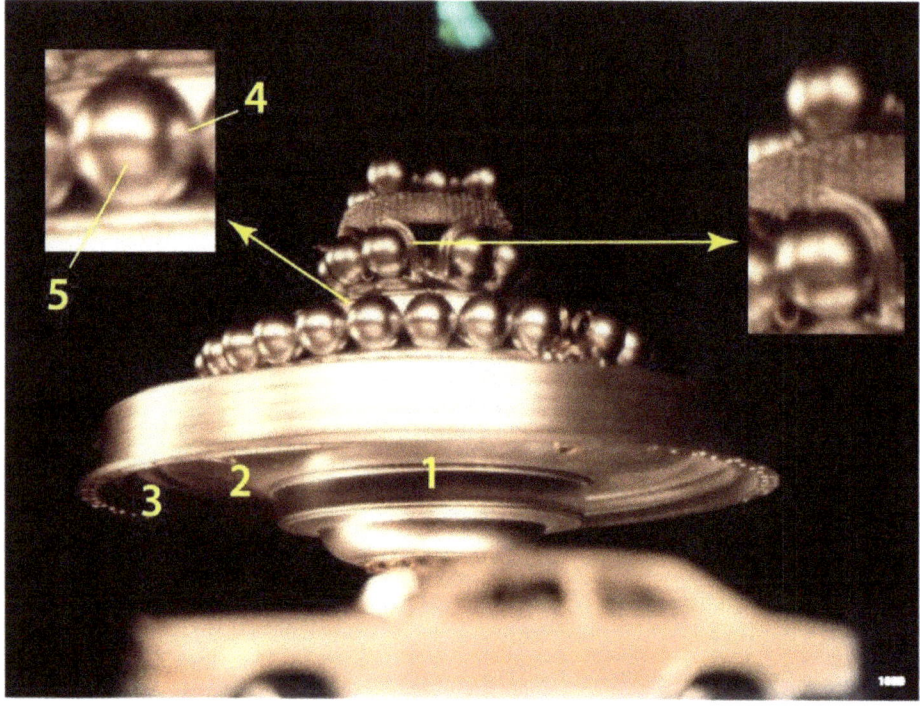

Figure 73 -   Photo #1000

Number **1** in Figure 73 is a significant reflection on the front of the WCUFO. If this is from an external source, it might be behind Meier. Two smaller reflections are on the left. So we have three sources of brightness if these are exterior lights.

A few things, however, do not gel if these are external lights. The car's side facing Meier's camera is not as bright as the WCUFO, even though the big light behind Meier is now in front of the car's left side and closer to it than the WCUFO. This light source produces no reflection on the car, and its general illumination seems the same as that in photo #999, which was caused by the street light. The light that "must be" behind Meier is not illuminating the car. Also, the top right zoomed WCUFO sphere shows three reflections with its central brightness at the right rather than the centre. A dark band occupies its centre. No correlation exists between the "reflections" on this sphere and the WCUFO base. Therefore familiar outside light sources cannot create them.

The lower spheres show reflections more compatible with the WCUFO base. Comparing the three levels of spheres, however, they do not match. The WCUFO base has no source of light on the right side; the entire right side of the base sections is dark. Why then does the top sphere show brightness on the right side? The other spheres are behind this sphere, so this reflection cannot be from those other spheres. Bright area **5** is a reflection from the WCUFO base. Brightness **4** comes from the contiguous sphere.

Also, the bottom left of this Figure 73 WCUFO shows illumination. Are there lights on the ground? Even if this were a scale model, bright reflections from potential studio lights do not match up. The most plausible explanation to date is that this WCUFO radiates light, as Meier claims, in a way that we do not yet fully understand.

Figure 74 shows photos #921 (top) and #870 (bottom), both taken on the same night as the previous photos. In full colour, the top photo shows a bluish-white street light on the right, noticeably illuminating the colour of the leaves. What lights not illuminating any other object in the scene could give this WCUFO its golden colour if the orange is not a plasma sheath?

The arrow indicates a bright spot on the cylindrical surface of the undercarriage. Where is the source of this light? Is it from the lamp on the right? If so, why is it golden and not bluish-green since the WCUFO is silver in daylight and the lamp is a whitish blue-green? Why, also, does the right basal side of the WCUFO not show the same light reflection, or the broad vertical base above it? The street light is not interacting with the WCUFO. Also, in this photo under a more detailed analysis, at least one car is found on the bottom right side. Why is this car not illuminated by the same lights that illuminate the WCUFO? If the WCUFO radiates light, why does it not illuminate the car or the other surroundings? These details strongly suggest this WCUFO is using a cloaking

They are Here

device like the one in photo #873 that was studied earlier. These questions also apply to the bottom photo in Figure 74.

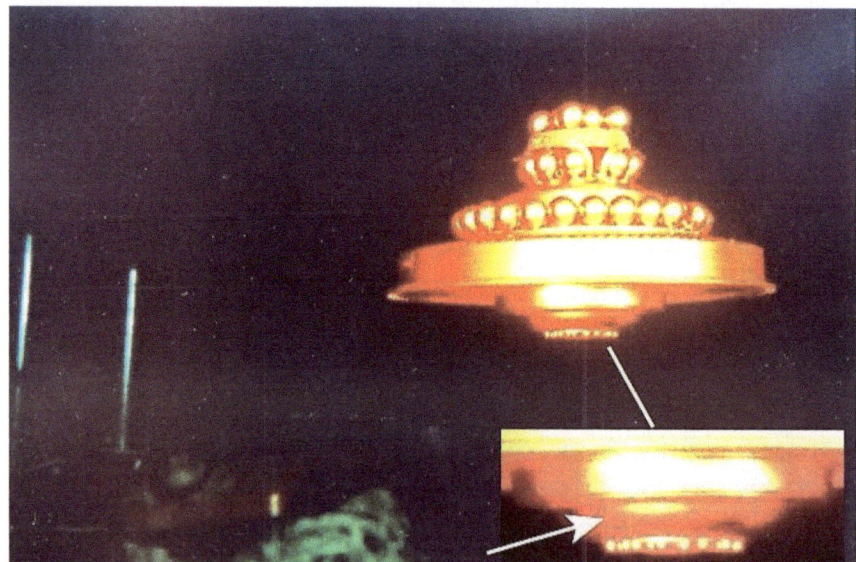

Figure 74 - Top: photo #921. Bottom: photo #870. Note the orange plasma sheath around the WCUFO

In photo #921 (Figure 74), the pole colours on the left and the bush colour at the bottom show an environment illuminated by a bluish light. Why do the golden colour lights illuminating this WCUFO not illuminate objects like the poles with the same golden colour? The plasma is *dull* energy that does not illuminate surroundings, but it might be picked up more significantly by the camera. Note the bright orange plasma plume around the WCUFO in photo #870. Judging by the light areas on this WCUFO base there are two sources of light, one just left and one just right of centre. Why does the inside penultimate lowest central base below the cylindrical surface show a thin horizontal golden brightness a bit to the left but not on the right? (See bottom corner detail.)

Furthermore, the 24-odd crystals on the very bottom of the WCUFO are either reflecting or emitting light brightly, perhaps intermittently since not all appear to be shining, yet never in any other photos. What creates this entire semi-circle of lights? These cannot be reflections of an entire semi-circle of external lights.

Better resolution copies of the photos are required to find even more details in the darker areas, such as vehicles parked nearby.

To conclude that external light sources illuminate the WCUFO or a UFO model leaves unexplained the many inconsistencies and incongruences in these photos. Even if the photos were the product of double exposure tricks, the "reflections" remain unexplained. A model illuminated by external studio lights does not in any way explain these inconsistencies and incongruences.

Finally, photo #871 (Figure 75) shows further inconsistencies. Here too is a green tree or bush with distinguishable leaves, this time below the WCUFO. Due to its good focus and the camera lens stuck on infinity focus, the tree or bush and the WCUFO must be a fair distance from the camera, assuming no double exposure. Why then is the WCUFO golden but the tree leaves have no trace of orange or yellow?

The zoomed sphere shows two sources of light, one on the left and one on the right. The WCUFO base shows "reflections" of these two lights, and the photo shows them reflected on the bottom concave cylindrical base (Figure 75 bottom boxed details). We explain shortly why these lights should not be there if they are reflections from external sources. With a concave cylindrical surface rather than a convex one like we have been looking at before, the reflections must flip horizontally. Details in the inserted bottom boxes of Figure 75 show rear "reflections" that should not

They are Here

be there. The one in the righthand box must come from a light source on the left, and behind Meier; but the extended bottom core must block such a light. So, no brightness from a left side light source should be there. The box on the left shows a similar "reflection." In this case, a single spot of light. How is it formed there? A righthand light source cannot illuminate this section of the WCUFO because similarly, the central core extended below must block it. Furthermore, why is it just a single spot of light, or a light extending somewhat around the back of the concave perimeter yet hidden by the extended central core? Again, even a UFO scale model illuminated by studio or external lights cannot explain these incongruences.

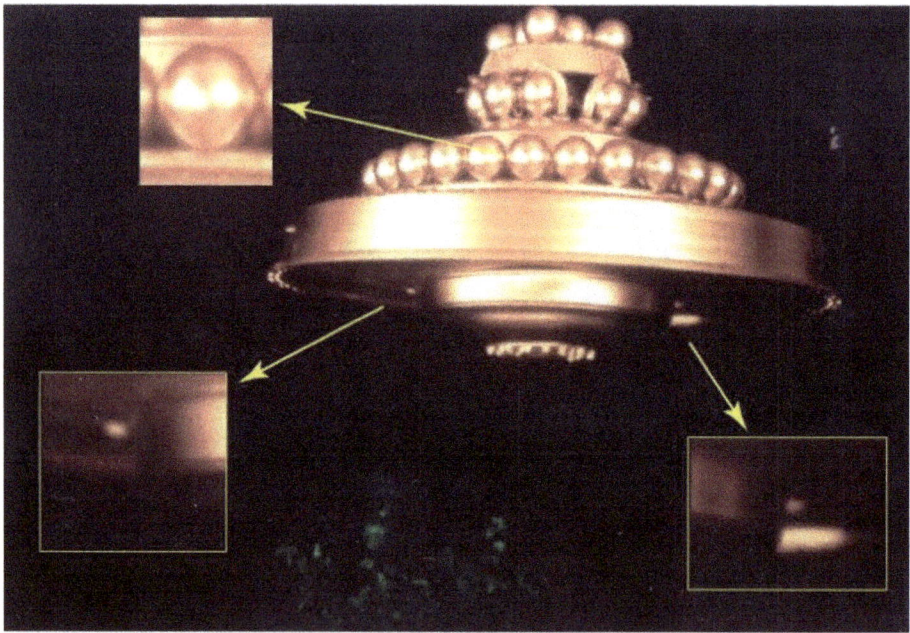

Figure 75 - Photo #871 details.

Like the previously analysed photo #873, these photos present insurmountable difficulties when trying to explain the WCUFO "reflections" as caused by external sources of light, either of another WCUFO or a scaled UFO model with or without a double-exposure trick. If the light is coming from the object itself, by radiating light instead of reflecting light, making a UFO scale model like this that radiates light would be most challenging to accomplish; and why, given bright model lights, are the WCUFO surroundings not illuminated? We have to conclude the genuine possibility of a cloaking device in use here.

We, therefore, conclude Meier's claim that the WCUFO shines at night by *its* energy or light sources not by reflecting external light bulb sources, is the best explanation of the photographic details. However, to be logically and scientifically sound, we do not know how the craft does this.

It is possible, therefore, to explain this 7-metre diameter object flying freely around various locations in the Swiss countryside with a surrounding sheath of ionising radiated plasma energy shining towards Meier's camera in the darkness, as most probably using a screening mechanism to occult itself from curious observers. But why is the WCUFO orange when in daylight it is silver?

Why is the WCUFO a golden orange colour?

While a simple explanation for the WCUFO's golden colour is its emitting or reflecting orange light, it is more likely, as explained earlier, due to atmospheric ionisation or a sheath of plasma close to the WCUFO's surface which creates an external orange light effect. Hill again says this is typical of UFOs at night, and even at times in the day time. However, in the daytime, the dull orange is doubtless invariably outshone by the bright blue-white light of day. Among the many reports of UFOs, reports of orange ones are not uncommon. But why orange?

As a reminder, Hill explains that the "ionized atmosphere, or plasma, ... gives them [UFOs] nighttime illumination in red, orange, yellow, green, blue, or white, and often gives them a mist-shrouded appearance" (page 27). Essentially, this explains the red and orange colours associated with this nighttime WCUFO. The "mist-shrouded appearance" is often not present, especially when the craft is stationary; when it is at a state of low energy activation.

Hill further explains that the colour seen relates to the amount of energy activation by the UFO. Basically, "[b]lue, white, and blue-white are the common colours at high-power operation". At the same time, "red and orange correspond to the least energy" and "are also the two most common colours associated with UFO low-power operation, such as hovering or low-power maneuvres" (page 62).

So, this nighttime WCUFO was in a low state of energy activation flying slowly or hovering, and indeed at this time, it must have been judging from its very close proximity to the ground at just a couple of metres below as Figures 64 and 65 attest.

It is important to note, as other UFO witnesses have pointed out, that "the illumination comes directly from the air and not from the vehicle surface" (Hill page 64). What this means is the WCUFO is not emitting orange light per se, it may be emitting light, but the orange colour is due to the low energy excitation of the ionised thin atmosphere around the WCUFO. Plasma brightness is "the result of an increase in the activation power that the UFO puts out" (page 64), but dull colours like the dull orange we see in the nighttime WCUFO suggest, again, lower power activation of the WCUFO. The WCUFOs here seem to be in easy-cruise or even *hover* mode, and when in Meier's courtyard, they are indeed hovering.

A difference in the energy spectrums of the orange and wine-red ionisation must exist because they exhibit different characteristics: the orange snuggly surrounding the whole craft, while the wine-red ionisation drifts upward or remains upward as the craft has slowly descended.

Finally, regarding the plasma sheath colours, Hill notes, "the UFO radiation causing the brightness is an integral part of the power system". In contrast "the observed atmospheric colours are a by-product of the power plant radiation" (page 65). Presumably, therefore, as mentioned earlier, separate power sources produce the red and orange colours associated with this nighttime WCUFO.

Probably the overall orange relates to a general lower surface power source or, as previously mentioned, the low power spherical *swinging wave accumulators* which cover the outside body from the lower to the upper level, and the many activated lights discussed throughout this chapter.

All this makes it very reasonable, if not inevitable to conclude that the WCUFO may well be a flying machine made by a technology currently unavailable on Earth, and therefore piloted

WCUFO at Night Analysis

onboard or remotely, as Meier says, by people of an extraterrestrial culture.

Do WCUFO spheres change their shapes?

Not all the spheres of this WCUFO in photo #873 look like pure spheres. Is this an optical illusion or a real change in shape? See Figure 76 showing the details of the deformed-looking spheres.

Using the Blender computer-generated model of this WCUFO duplicated this effect precisely as happens in the computer photo reproductions. The reflections on the spheres create an optical illusion that makes them look like deformed spheres. See Figure 77.

Thus, this visual illusion relates to the reflections between and among the contiguous spheres under high contrast lighting. This effect is not present in typical daylight situations. So the WCUFO spheres do not change shape.

Figure 76 -   Sphere deformations?

They are Here

Figure 77 - Model sphere deformations created in Blender. Deformation is an optical illusion caused by reflections on contiguous spheres under high contrast conditions.

Summary of WCUFO: Analysis and Conclusions

After conducting several tests and calculations on several WCUFO photos, we conclude the following:

- There are several WCUFOs, not just one. One is around 3.5 metres in diameter. This one was hovering above Meier's parking lot, then above the treetops, and then hovering stationary very close to a Norway Spruce tree.

- Another WCUFO visible in night photos, measures around 7 metres in diameter, and its dull orange colour is almost certainly due to the thin ionised atmosphere immediately around it. This thin, dull golden light is presumably overridden in daylight by the bright blue-white light of day. We estimate the WCUFO size by comparing it with a car in front, and a pole in the terrain below it.

- The WCUFO flying nearly above Meier's trailer might be either the 3.5-metre or the 7-metre WCUFO, while Meier says it is the 7-metre WCUFO.

- The 7-metre WCUFO (photo #873) extends its central core upwards. The same 7-metre WCUFO close to a Mercedes Benz does not, or did not extend its central core.

- There is no possibility that this craft is a small 55 cm diameter model made from household items like a trash-can lid and Christmas tree balls. Mathematical and geometrical calculations show that is patently impossible.

- Sceptics base their comments and claims, that this object is a small scale model made from household items, on intuitive perception, guesswork, and making a model that looks very much like the WCUFO minus important details; not scientific analysis and testing. "If it looks fake it must be fake", is not a valid scientific standpoint. "Looks fake" is a subjective opinion.

- It is surprising how similar some parts of this UFO are to household items. We think it not a coincidence. Were these household items designed and constructed after the WCUFO photographs reached public attention? Some household items look similar to some aspects of the WCUFO. They are not identical, however, note the top platform that is similar to an upturned flower pot tray but far bigger.

- The WCUFO has abundant exotic details, like coloured crystals, coloured lenses and delicate engraving on metal parts that prove troublesome to duplicate even in a small model. Most importantly, we know these WCUFOs are about 3.5 m in diameter or more.

- The WCUFO is unlike other UFOs photographed by Meier or other flying saucer-like UFOs seen in photos worldwide. It appears to be a big flying object designed and constructed to operate in the earth's magnetic or gravitational field. The absence of its aerodynamic top cover, if it had one, reveals all its interior details. Does the WCUFO give clues on how to construct a flying machine like this?

- Could the spheres be small engines providing this UFO with stability as it floats and moves, perhaps by interacting with our magnetic or gravitational field? Meier says the Plejaren call these spheres "swinging wave accumulators." We might call them high- or subtle-energy accumulators or resonators. If so, perhaps their low energy accumulation or activity is responsible for the dull nighttime golden-orange plasma glow around the WCUFO surface. Furthermore, many night lights shine from within various sections of the spheres and craft, any or all of which could be emitting low energy fields responsible for the golden-orange low energy plasma glow. The WCUFO gives rise to more intriguing questions than it provides answers.

- The bright golden-orange nighttime WCUFO colour is not explained by external lights, reflections, like street lights if it is a big object or by studio lamps, if it is a small model. Instead, a thin sheath of *low energy* plasma

Summary of WCUFO: Analysis and Conclusions

closely surrounding the craft's surface explains it. The most logical explanation for the lights on the WCUFO at night is that the craft *emits its* light by an unknown mechanism, the light becoming golden as it passes through the ionised atmosphere directly surrounding the craft.

- Photo #873 shows a somewhat inexplicable surrounding environment with a dark terrain and a pole casting a single shadow below the very bright WCUFO. The solitary yard light at Meier's house explains the single shadow since more external lights would create more pole shadows. Meier's explanation for this is that the WCUFO emits light rather than reflects it, and the craft has a cloaking system blocking light escaping it except for a small opening, "doorway" or "passage" of light and visibility to Meier's camera and head allowing him to take photos.

- Other WCUFO night photos like #870, #871, and #921 show a dark environment around the craft. But if external lights cause the bright reflections on the spheres, it is inexplicable why everything around the WCUFO is so obscure and dark. In one photo, at least one car is visible in the darkness, but it reflects no lights from any external sources. A single bluefish lamp is in the vicinity, but it is very faint and does not illuminate the surroundings. Attempting to explain this photo as a little model and a double exposure trick is illogical because the reflections follow no logical pattern caused by studio lights. Again, as in photo #873, the most likely explanation is that the WCUFO is a significant object radiating light and that it appears to have a cloaking device blocking any emitted light, except through a narrow "window", "doorway", or passage of light towards Meier's camera, face, and head.

- There is no reasonable scale model explanation for the pink or wine-red plume around the WCUFO in photo #873. The plume is most likely a not uncommon UFO plasma field, and the underlying technical details behind it are discussed earlier in this chapter. (For more information see Hill's book *Unconventional Flying Objects*.)

They are Here

- There is no evidence suggesting a photographic trick, like double exposure, photomontage, or the use of a small model.

- Surprisingly, abundant clues in these photographs have remained hidden for decades until recently. Doubtless, curious researchers who patiently research and employ scientific processes on these unique photographs will discover more in the future.

- The Meier case evidence is not only "too good to be true". It is "too good to be false" and so *is* true.

- By experimenting with this evidence, readers can confirm our findings. However, with the arrival of the required worldwide UFO controversy in the twentieth century, sceptics and debunkers may persist in debating and rejecting this evidence; assuming they even dare to look through the telescope to see the proof.

Part II

An Investigation into *The Pendulum UFO*

Introduction

This 18 March 1975 movie is the first of Meier's films, shot on a cold and hazy, or snowy, evening. It is popularly known as both *The Pendulum UFO* film and *A UFO Circling a Tree*. Part II of this book is an updated investigation of this film based on new evidence and new findings.

In 2014 we published results of an initial investigation that concluded attempts to explain this flying object in Meier's first film as a little model hung from a cord faced unsolvable problems. We found the craft most likely to be a sizeable flying disc somewhat imitating pendulum movements. It is too difficult to explain this object as a simple model in a pendulum, due mainly to significant variations in the pendulum periods.

Now, after viewing a full-length video and investigating different versions of the same movie, we can safely conclude it is a real and sizeable flying disc. Anyone still thinking this is a little model is not looking at the whole evidence. Instead of calling this movie *The Pendulum UFO*, however, it would be more accurate to now call it *The Forced Pendulum Beamship*, beamship being the term the Plejaren ETs use to describe this and other types of their spaceships.

Several sceptics have tried to reproduce this film, or provide some explanations, but in their attempts, they have failed by ignoring some of the film's essential details. We have found that in the Meier case, what appears quite simple to explain on first impressions is not simple at all and requires a detailed analysis. As with the WCUFO analyses, clues reveal *The Pendulum UFO* case is not as simple as some might first think. The evidence here, as in other Meier evidence, leaves room for some doubt and an overly simple initial assessment. After looking in more detail, however, hidden clues arise showing this is no simple trick, but rather a sophisticated attempt at creating a controversy surrounding UFOs.

Our introduction to this book explained how the Meier case has contributed to the worldwide UFO controversy, and indeed more so than any other UFO case. A fact which even the most hardened sceptic would probably not deny. Given humanity's broad spectrum of rigid, intolerant beliefs and ideas, especially on such subjects as humanity, religion, life, space, and ETs, a UFO controversy was probably considered necessary. Why? To help us develop the changes in our thinking needed to enable safe passage through the immense challenges we now face; a shift in thinking necessary for us to become a better and space-faring future Earth humanity.

Can we demonstrate, in straightforward terms, that this is a big flying disc and not a small model hung from a cord dancing around a little tree? If we look at the evidence we find:

Figure 78 - The beamship, tree, and houses show the same level of washed-out bluish-grey tones, while the foreground grass at the bottom reveals darker and more vibrant colours. These picture data tell us this is not a little tree close to a little UFO model at only 17 metres from the camera.

- A snapshot image from the film with enhanced brightness and contrast shows the tree, houses, and beamship far away, Figure 78. It is well-known that Rayleigh light scattering causes distant objects to lose colour intensity, and dark tones, turning more distant

Introduction

objects a lighter bluish-grey. At the same time, near objects present more vivid, intense, and darker colours (Zawischa "Light Scatter"). In Figure 78, we notice much lighter grey tones on the distant house, tree, and beamship compared with the nearby dark green grass at the bottom of the image. If the tree is small and just 17 m from the camera rather than a 20 m tall tree at 385 m, it would appear sharper, darker, in better focus, and more vivid in colour, like the foreground grass.

- The beamship moves the treetop without touching it, which is so difficult to achieve with a little tree that no one has ever managed it. The tree's movement indicates the beamship is interacting with it, confirming the beamship's proximity to the distant tree and house.

- The size of the house, tree and beamship, confirms the size of this flying disc is around seven metres, as Meier reported.

- Other aspects of this investigation give additional details on this beamship's flying capabilities that are not found or created in a hanging model.

Changes from the 2014 investigation:

In the 2014 investigation, we studied available copies of *The Pendulum UFO* film. One such copy was a projected image from a video published by FIGU on their Switzerland website. Another was from the "FULL Billy Meier-1985 Beamship - The Movie Footage" (Yaoi). Now, more copies are accessible, like the *YouTube* documentary "Pendulum UFO - Demonstrationsflüge / Demonstration flights (detail FIGU)" (Nov. 2019) showing the full-film sequence in better quality and greater detail. In the next section, we detail the origin of this copy and provide some technical information on PAL and NTSC videos.

This new analysis gives more precise information confirming this flying object is a large disc and reveals further previously unknown details. The main changes introduced with this new analysis are:

They are Here

- There are three beamship jumps, not just the two the previous investigation found.

- The 2014 investigation concluded the jumps take around $1/10^{th}$ of a second, and that they happened in a gradual transition indicated in three consecutive frames. Now we know the jumps are faster, changing from one frame to another, in less than $1/24^{th}$ of a second.

- We now know that the video used for our 2014 analysis was a VHS PAL format copy of the original video, combined with a projection and recording in a 30 fps video which caused a "smoothing" effect and a white band during the jumps to happen in at least three consecutive frames. Now we know there is only one white band during each jump.

- A detail like a VHS copy distortion, projected as a bright horizontal line crossing the beamship image, is not a distortion, but a genuine unknown manifestation that occurs four times. We call these manifestations "pulses" and think it not coincidental that this bright curved line always crosses the beamship images. It suggests the beamship is emitting some pulse that affects the film surface, and not an intentionally faked scratch on the negative.

- A flip happens 16 seconds after the last jump. This beamship flip occurs almost instantly, in less than $1/24^{th}$ of a second, a manoeuvre no model or known sizeable flying craft can perform.

- Additional features like bright flares are found that could be caused by electrostatic charges inside the camera.

- The beamship in this film, as in others recorded by Meier, presents a wobbling movement incompatible with a model hanging from a cord or wire. A model's point of rotation or support is where the wire connects to the top of the model. In the film, however, it is centred inside the beamship near its base.

Evidence from Four Videos

Introduction

Several copies have been made from the original film, mostly in the 1970s when it seemed to have passed not infrequently between the hands of researchers, visitors, and family members. While some researchers think the original film is the only acceptable evidence to work with, the several extant copies can be considered a valuable piece of research material. Old films may last 20 to 50 years maintained in ideal environmental conditions of temperature and humidity. Meier filmed his movie in 1975, 44 years ago. The original film must have severely deteriorated by now unless stored explicitly at optimum temperatures and humidity. Old films lose their colour, turning yellowish, the film shrinks, and the emulsion breaks off the cellulose, eventually reducing the film to bubbles or dust. How can we preserve an old film as close as possible to its pristine condition if we lack a perfect environment? Our best option might be to make copies. Various people did this, and several different video copies recorded at different times now provide the cumulative evidence of *the Pendulum UFO*.

Of course, we would all like the original film of the beamship demonstrations. So, where is it? The short answer is: no one knows. Not even Meier knows where it is. Presumably, wherever it is, unless filed safely in some secret science lab, by now it must have severely deteriorated or even been destroyed. This section begins by presenting what we currently know of the history surrounding the film.

The best evidence is the several copies we find showing original details and the evolution of the film's deterioration. For example, the movie by Stevens and Elders shows a short segment of *The Pendulum UFO* film that includes the final phase as the beamship ascends high into the sky (*Contact* ...1982). In this copy, we do not see the few burned frames present in later copies. Also, this short clip does not show the scratches or dust particles associated with later copies. Furthermore, the images of the house, the treetop movement and beamship are seen more clearly in this copy.

They are Here

Figure 79 shows different aspects of deterioration in the film marked **1** to **4**. Marked **1** is the most common: dust. Dust particles liberally pepper specific frames of some film copies. Number **2** are ubiquitous vertical scratches present in an old film. During film projection dust particles get trapped in the projector, scratching the film surface as it moves through the projector. As a result, we see several vertical lines undulating along the film's surface. Number **3** shows fibres. Like dust, microfibres also float in the air, eventually fall over the film surface, and remain there. Number **4** is the most critical: and manifests the film emulsion containing the images peeling or falling off the film surface.

There is more damage not indicated in this figure, like non-vertical scratches and burned frames. During movie projection, if the camera-man stops or pauses the film before the end, the bulb may be off, but it is still so hot that it burns the frame in front of it. During the continuous projection of the film, the emulsion does not burn, but if the movie stops, it burns a frame.

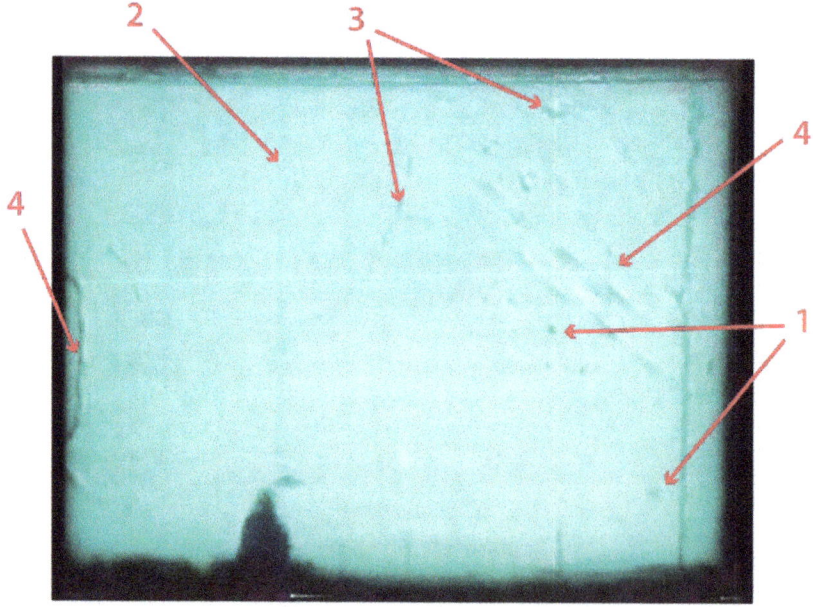

Figure 79 - Types of film deterioration: **1** Dust, **2** vertical scratches, **3** microfibres, **4** emulsion peeling, or falling off the film surface.

Evidence from Four Videos

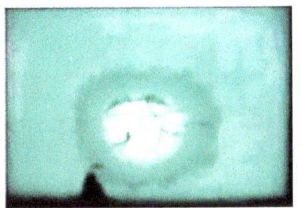  

Figure 80 -   Burned frames in the film.

These damages, however, can be used to advantage. They help differentiate each frame in the original movie and indicate whether the copy is an earlier or later version. For example, as detailed later, a PAL format movie produces two interlaced images. Seeing the same piece of dust in two consecutive frames in the PAL format copy (two interlaced images) shows not one frame at a time from the original film but an overlap of images. Knowing the format of the copy is essential to determine the nature of the original movie.

History and evolution of the film

What happened to the original film and its copies? We find some answers by reading the Foreword to the documentary "Demonstrationsflüge/Demonstration flights (IFOs, not UFOs)" on the FIGU YouTube channel:

Vorwort	**Foreword**
Diese Edition des Films ‹Demonstrationsflüge› ist neu geschnitten und zusammengestellt und unterscheidet sich im Wesentlichen von der bis jetzt (April 2005) von der FIGU zum Verkauf angebotenen Vorgängerversion auf VHS-Kassette.	This version of the movie „Demonstrationsflüge" (demonstration flights) has been newly cut and assembled, and it substantially differs from the old version on a VHS cassette, which has been offered for purchase until now (April 2005).
Leider wurde der im Original zwei Stunden dauernde 8mm-Film von Billys Ex-Frau entwendet und aus Verleumdungs- und Profitgründen an einen Mann in Deutschland verkauft.	Unfortunately, the original, two-hours long 8mm film was pilfered by Billy's ex-wife and, for reasons of defamation and profit, was sold to a man in Germany.
Der Film wurde von diesem total verschnitten und verschandelt, so dass Teile rückwärts liefen und die	He totally cut and mutilated the film, from which resulted that some parts were running reverse, and that the

German	English
Ausschnitte nicht mehr in der Reihenfolge stimmten.	cuttings were not in [the] correct order anymore.
In mühsamer Recherche wurde der verschnittene Film von Michael Hesemann, Chefredaktor des Magazin 2000, wieder aufgetrieben und auf VHS-Kassette kopiert.	After arduous enquiry, the mutilated film was found, and copied onto a VHS cassette, by Michael Hesemann, the chief editor of *Magazin 2000*.
Von dieser Kopie ist das Material aufbereitet und mit Billys Hilfe und Einverständnis neu geordnet, geschnitten und vertont worden.	Based on this copy, the material has been edited and with Billy's assistance and permission newly sorted, cut and sound added.
Bei der Verschandelung des Films wurden auch die Kommentare von Billy verstümmelt und sind zum Teil abgeschnitten und unvollständig.	Through the mutilation of the film Billy's comments were garbled (became?) and in parts were cut and incomplete.
Das Neubesprechen würde aber viel zu viel Zeit in Anspruch nehmen, so dass darauf verzichtet wurde.	Since a new narration would require too much time, it was omitted.
Diese neue Version dauert leider nur noch ca. eine Stunde, ist aber hoffentlich für die Zuschauer trotzdem noch interessant und aufschlussreich.	This new version only lasts about one hour, unfortunately, but hopefully is still interesting and instructive for the viewer.
Wir bitten um Verständnis und verbleiben mit freundlichen Grüssen.	Thank you for your understanding.
FIGU	Sincerely yours,
	FIGU

Film details, later corroborated by Meier, are also in Wendelle Steven's book *UFO Contact from the Pleiades – A supplementary investigation report* (page 237). We also asked Meier personally, through Christian Frehner, regarding the evolution and events surrounding this film. Thanks to their replies, we have been able to reconstruct its history as follows:

- Initially, the various 8 mm films Meier took of the beamships were on different film rolls. These were assembled into one film/footage by a professor whose name Meier does not remember at this time and who visited Meier several times in Hinwil. Meier did not have a film laboratory, a darkroom, or skill in editing or developing photos or videos, so presumably, he received help from individual experts.

Evidence from Four Videos

- Later, the film was received by Mr Schmid and another man who took care of Meier's photos, sent negatives to the laboratory and performed other professional assistance.

- This 8 mm assemblage was then used as the original to order 8 mm copies that were subsequently sold to interested persons at the SSSC. Most probably, Wendelle Stevens had one such copy described in his second book (*UFO Contact...A supplementary...*). When Denver reporters visited him at his home, despite all attempts to record it they mysteriously could not do so (page 536). We might also assume that the segment seen in the movie *Contact* is from a Steven's or another copy acquired by Lee and Brit Elders.

- Stevens described in his book how TV station reporters from Munich visited Meier to make a "Special" Austrian TV program. They convinced Meier to give them his long, originally composed film, to make a videotape copy. They took the original film with them, promising to return it soon.

- It took a long time for Meier and FIGU to recover the film. When they received the lengthy film, in its original reel, they put it in a safe place, without conducting a detailed check to verify the status of the film. They assumed they had received the original film because it came in the original reel.

- In 1979, Nippon TV personnel visited Meier and produced a special documentary on the Meier case. Meier was showing them the long film that had been stored in a safe place when he noticed some pieces were missing. After checking the whole film, they discovered that not only were some clips lost but also the existing ones were in a different sequence, some even in reverse order. There were no splice marks on the original film that should have been there, between each clip. The realisation dawned that the film in its original reel was a copy, not the original. Some spliced segments were out of sequence, and a few segments ran backwards. Presumably, someone had copied this edited film, and Meier received the copy in the original reel. Nippon TV interviewed Meier and recorded a 3/4" videotape in NTSC format, from the projected copy of Meier's film. *The pendulum UFO* video we now see in the Nippon TV documentary is, therefore, a recording of a projection of a copy of Meier's film, not the original.

- The Austrian TV station claimed they returned the original film to Meier, not a copy.

They are Here

- Today, it is not clear whether Meier received the original film from the Austrian reporters as they claimed, or a copy, or whether someone replaced the original film with a copy during transit between Austria and Switzerland.

- Some years later, Meier found that "Unfortunately, the original two-hours [sic] long 8mm film was pilfered by Billy's ex-wife and, for reasons of defamation and profit, was sold to a man in Germany" (FIGU, Foreword, *Demonstrationsflüge / Demonstration flights*).

- The documentary *Demonstrationsflüge / Demonstration flights* was another copy created from a recovered copy – of the original? – by somebody in Germany.

- Currently, several copies of the film must exist in various collectors' hands, and, as stated earlier, the location of the original film remains unknown, even to FIGU members and Meier. However, since so much time has passed subsequent film deterioration is inevitable. It must be in poor condition. Right now, the best evidence available lies in the various common aspects and differences within the copies that we can identify and analyse to see what they reveal.

Evidence from Four Videos

The four videos and sources used in this investigation

Four different available videos are used for this, our second investigation. Perhaps, and hopefully, more videos or good copies of the original will surface in the future. Every time we look again at the videos or *new* ones surfacing, surprising details reveal themselves. As mentioned, several copies taken at different times is arguably better now than having the original film due to the latter's expected severe deterioration in quality.

1. *Contact* movie film

Source: *Contact – 'Billy' Eduard A. Meier Documentary by Wendelle Stevens* (1982). The famous documentary produced by Brit and Lee Elders.

Format: MP4, at 25 fps. Low resolution.

Link: https://youtu.be/gGCuLIVxxQw

Figure 81 - Snapshot from the movie *Contact*.

They are Here

The *Contact* documentary shows this small fragment. Although lacking quality, it very clearly shows how the beamship moves the tree without touching it. The clip seems to be from an early copy of the original film. Noticeably, there are no burned frames during the last phase when the beamship ascends in a spiral pendulum movement. Later film copies show burned frames in this phase, yet their beamship, house, and tree details are comparatively useful. This clip's disadvantage is that it is neither a full-frame nor a full-length copy of the demonstration. However, a zoomed image gives excellent details of the tree's movement.

2. Video previously on FIGU website

Source: A copy made with a screen capture tool from a video previously available on the FIGU website, and the video used in our first investigation into *The Pendulum UFO* in 2014.

Format: MP4, at 24 fps. Low resolution.

Link: https://youtu.be/K_HnDz4KY6k

This video has the advantage of being in full-frame and reveals several details in the image but lacks any zoom. The quality is not particularly good perhaps for having been subject to several previous transformations, from the super 8 mm film to a PAL format and then to a 30 fps. However, it shows most of the movement phases excellently, enabling pendulum period measurements and their comparisons to other video versions. A disadvantage of this video is that it does not show the final part of the film that includes the third jump, the abrupt beamship flip, and its final ascending spiral movement.

For comparison, Figure 82 is a snapshot from this video at a similar moment in the film to the *Contact* image in Figure 81.

Evidence from Four Videos

Figure 82 - Snapshot from the video *Hinwil* that was available on the FIGU website.

3. Nippon TV documentary

Source: Filmed by Nippon TV during their documentary on the Billy Meier case in 1975.

Format: MP4 from NTSC, at 30 fps.

Link: https://youtu.be/WkQgwlPPLZM

This NTSC format video has interlaced frames. To understand the process of capturing or copying the original video images, we explain the relevant PAL and NTSC format characteristics. The advantage of this video is that it zooms on details like the "jumps". Based on Stevens' description in his book *UFO Contact from the Pleiades: A Supplementary Investigation Report* (pages 237 ~ 238), we can conclude the projected video is the copy Meier received. We found some frames at the beginning of *The Pendulum UFO* video cut out together with some frames at the end of the previous clip where Meier shows the same location with the houses minus the big tree. So this copy shows the same scenes, but with some frames cut out. Perhaps the beginning of *The Pendulum UFO* video was cut to omit the initial frames showing the oval pattern suggesting the start of the film roll. (Shown in Video 4 details.)

Figure 83 - Snapshot from the NTSC video in the Nippon TV documentary.

Another characteristic from this video is its backwards repetition, where a bright band at the bottom of the image appears. This bright band resulted from moving the VHS tape in the opposite direction. So, instead of the repetition showing the event sequence right turn, first jump and second jump, it shows the sequence, second jump, the first jump and the right turn at the end, and the beamship moves backwards. The operator stops the projector several times during the process of copying the entire movie, and this may well have caused the burning of a few frames at the end of this version of *The Pendulum UFO* film.

A significant disadvantage of this video is it lacks full-frame recording, so some details, like the white bands at the jumps, cannot be seen and compared with other videos.

For some reason, this video's time display indicates the video runs 8.5% faster than other videos.

Figure 83 shows an attempt to capture the beamship at a similar moment beyond the tree as in Figures 81 and 82.

Evidence from Four Videos

4. FIGU documentary:

"Pendulum UFO - Demonstrationsflüge / Demonstration flights (detail FIGU)" (Baseline)

Source: The full sequence of *The Pendulum UFO* from the FIGU documentary.

Format: MP4 from PAL, at 25 fps.

Link: https://youtu.be/Gzr3BRUhfy8

Figure 84 - PAL video snapshot from the FIGU documentary "Pendulum UFO – Demonstrationsflüge / Demonstration flights (detail FIGU)".

This (Baseline) video is the one used in this investigation. It is a copy of a PAL format VHS video and has interlaced frames just like the Nippon TV video. It contains the full sequence with all events. References to specific times from this video are in the format:

$$MM\text{-}SS\text{-}FF$$

They are Here

Where:

MM refers to minutes from the beginning. (0 to 59)

SS refers to seconds after the referred minutes. (0 to 59)

FF refers to the frame, from frame 00 to frame 24 (25 fps)

For example, in Figure 90, which visually lists all significant film events, the "Second pulse" happens at 02-13-08, meaning at 2 minutes, 13 seconds, and in the 8th frame after that time. Later we discuss in detail the mysterious pulses found in the film.

Finally, this film copy has more scenes, more frames and more details than the film Nippon TV analysed. For example, in this FIGU documentary film, some curious frames show an oval mark that could relate to the start of a blank 8 mm film cartridge that Meier loaded in his camera. Figure 85 shows two of these frames.

 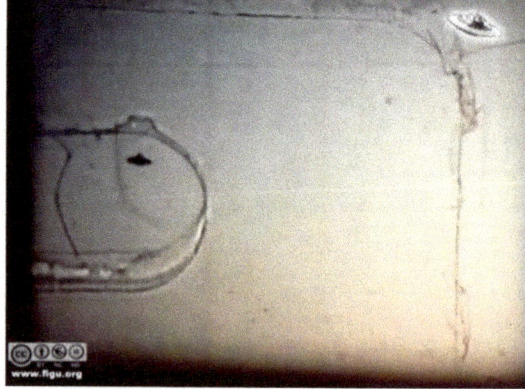

Figure 85 - Examples of "first" frames in a film roll. The oval marks might prevail at the beginning of every film roll. A few "bad" splicings of two rolls in the FIGU documentary film produce an occasional rectangular white band at the top of the image, as in the left side image. White marks are also present during the beamship "jumps", but they are different.

The right side image shows the oval mark clearly in the frame against a pink sunset background as Meier was filming the beamship that emitted mysterious flashes. It is clear his roll finished and that he loaded a new one to continue recording the demonstration. The left side starting frame in *The Pendulum UFO* film shows the same oval mark. The white band on the top is due to "bad" splicing of two clips. We find the same white band in some, but not all of the long film splices. We see later, another but different white band present during the beamship "jumps." This white band is discussed in detail in the section on beamship "jumps".

PAL and NTSC formats in *The Pendulum UFO* film

Two of the most remarkable Pendulum UFO videos available come from recordings of the original, or a copy, of the Super 8 mm projection. One is the Nippon TV documentary and the other the FIGU YouTube video (*Demonstrationsflüge / Demonstration flights*). Nippon TV recorded the video on NTSC format that produces 30 frames per second (fps). The FIGU documentary is in PAL format at 25 frames per second (fps). Both present clear examples of what is called the *interlaced video* effect.

It is crucial to understand the technical details of these formats to verify what was in the original Super 8mm film. For example, we may see two images in the PAL video, but they are from one single frame in the original film. Or we may see a *Venetian curtain* effect relating to the interlacing effect in the PAL or NTSC format, not the original film.

Interlacing

Old TVs displayed images at 30 fps in places where electricity was at 60 hertz, and in 25 fps where electricity was at 50 hertz. Someone watching a TV movie in a 25 fps TV, for example, sees the movement of an object as intermittent. Creating interlaced frames minimises this effect (see *Interlaced Video* Wikipedia).

An *interlaced video* means, for example, that a PAL format VHS camera takes two video images at 50 fps that combine into a final video frame in a video produced at 25 fps. Internally, the camera produces two images separated at 1/50 of a second and then combines them in one single shot, presented every 1/25 of a second. Each created image is in a *Venetian curtain* pattern that combined enables our brain to perceive as a smooth movement. (See Figure 86, illustration.)

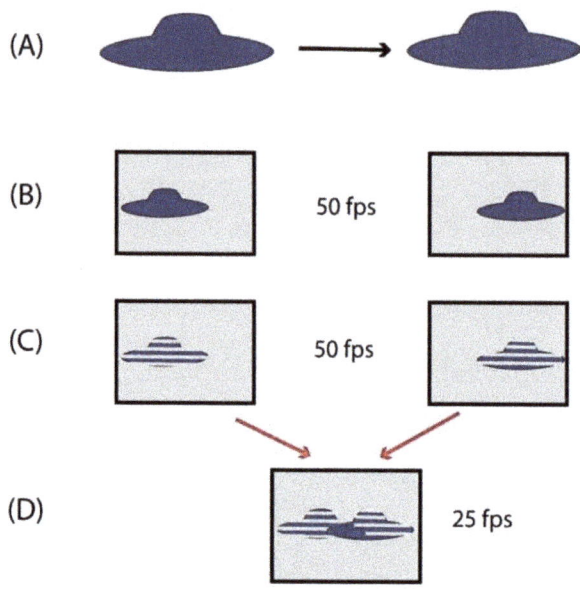

Figure 86 - Interlaced images in PAL format. Two images taken at 50 fps combine into one composed image at 25 fps, showing the *Venetian curtain* effect.

Suppose an object is moving rapidly from left to right (Figure 86 A). The VHS camera, using a PAL format, captures two different frames at 50 fps, every 1/50 of a second (Figure 86 B.) To combine them in one single image at 25 fps (Figure 86 D), it creates a *Venetian curtain* effect by having the *even* horizontal lines in one image and the *odd* horizontal lines in the other (Figure 86 C). The PAL format produces 625 horizontal lines.

The NTSC format also produces the same interlaced effect, but the frame rate is 60 fps producing a video at 30 fps with 525 horizontal lines.

Meier did not film the flying beamships with a PAL or NTSC format VHS camera. He used a Nalcom FTL 1000 Super 8 mm film camera. The VHS was used to record a projection of the original film or a screening of a pretty good 8 mm film copy, as seen in the Nippon TV documentary.

The YouTube FIGU documentary (Baseline) film used in this investigation is in PAL format, accessible here:

https://youtu.be/Gzr3BRUhfy8

Some digital video conversion systems are capable of de-interlacing a VHS film. It seems that when transferring the FIGU video into digital format, the de-interlacing capability worked only partially, so we see more extensive *Venetian curtain* lines. See the example in Figure 87.

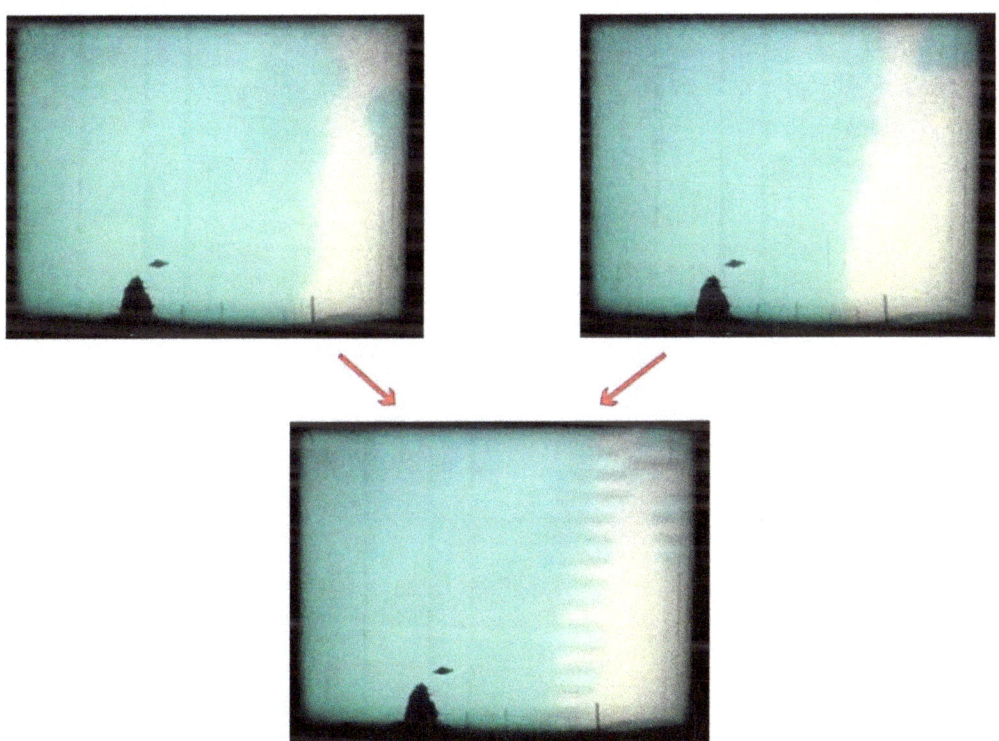

Figure 87 - Top: Two PAL format interlaced frames of *burning* on the film, produced probably by an electrostatic charge inside the camera. Bottom: An interlaced frame showing *Venetian curtain* lines.

The PAL video, as mentioned, captures two shots every 1/50 of a second, and interlaces them in one single shot. What might be

They are Here

the result of recording a PAL video from a projected Super 8 mm film? We find a fascinating pattern confirmed by the dust particles found on the film roll. Figure 88 shows the combination of a PAL video recording from a 24 fps Super 8 film. The top row (a) shows 24 frames from the original Super 8 film. Meier's camera can record at 18, 24, or 36 frames per second. We assume he used the standard 24 fps throughout the recording. If the frame rate would be different, the resulting pattern would very similar. The bottom row (c), shows the 25 fps in a PAL video. Each rectangle in (a) represents one frame of the original film, and in (c) we find the PAL copy.

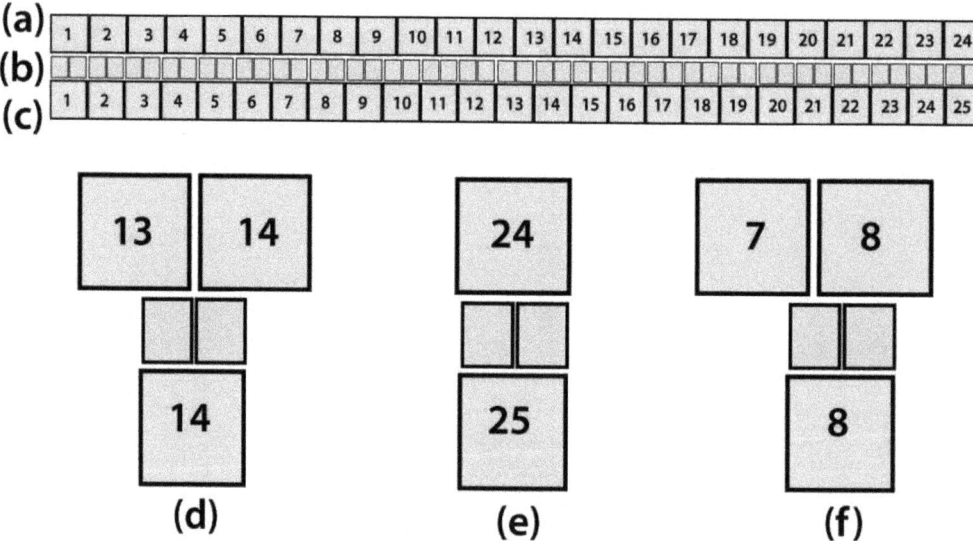

Figure 88 - Row (a) shows 24 Super 8 film frames in one second. Row (b) shows two interlaced frames at 1/50 that combine in pairs to create the PAL frames in row (c). Row (c) shows frames recorded on a PAL video at 25 fps.

Row (c), the PAL copy in Figure 88 shows how the two frames at 1/50 of a second combine in a *Venetian curtain* effect (Interlaced). Every frame on the PAL video combines two shots from the original film. The resulting pattern creates three different combinations. Combination (e) results in a PAL frame showing one single frame from the original Super 8 film; some are in both the PAL video (FIGU documentary) and the NTSC video (Nippon TV documentary). The NTSC format produces the same result in a pattern that repeats every 30 fps instead of 25 fps. This combination (e) is infrequent or lower in probability in the VHS videos than combination (f). In combination (f) the PAL frame shows two interlaced images from two different frames in the original Super 8 film. The combination percentages vary. It could, for example, be 10% intensity from one frame of the Super 8 film combined with a 90% intensity of the next Super 8 film frame image. Combination (d) is the same as the combination (f), but the percentage is 50% with 50% instead of 10% with 90%.

What usually results from this is a created combination of two frames or images from the Super 8 film. It is easy to see this effect "combine" the dust particles from the original film in the PAL and NTSC video recordings. Figure 89 illustrates the process. The dust particles, or any detail in the original film, are represented by capital letters.

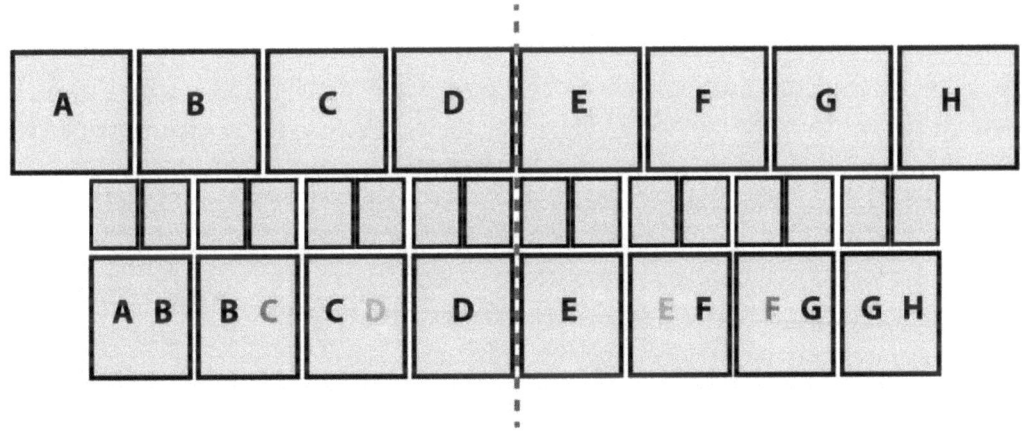

Figure 89 - Resulting interlaced images or frames showing dust particle combinations from two frames in the original film. Top row: each letter represents a unique dust particle or frame detail. Bottom row: the resulting combination in the PAL or NTSC format.

In Figure 89 diagram illustrates how most PAL format and NTSC format frames show a combination of two images. For example, dust particles marked C and D are unique in different frames on the original film, but the PAL or NTSC format shows them together, C with D looking faint in this example.

We can conclude that in the PAL and NTSC videos:

- Most of the PAL or NTSC frames are a combination of two frames.

- Single detail duplication in a single frame of the original movie occurs with high probability. In rare cases, only one unique image occurs with no duplicated image (non-interlaced images).

- We show momentarily how the beamship "jumps" occurred very rapidly, in less than 1/24 of a second. The PAL and NTSC formats show these "jumps" in two frames, but they happened in only one original frame.

- This pattern is easy to confirm in the Nippon TV NTSC video but not in the FIGU documentary PAL format because the latter was partially de-interlaced, making the *Venetian curtain* effect lines extensive. So it is challenging to spot tiny dust particles in some frames on the PAL video.

As noted earlier, it is vital to know which videos we are looking at and the laws governing their images. Perhaps in the future, due to the original video in Super 8 format coming to light assuming it still exists, or at least a good copy, a frame by frame analysis will be possible. However, even without the original film, high-quality evidence is available in the different copies made at different times; and together they may be better than the original movie that today may well be in a dilapidated condition.

As discussed in the Introduction of this book, the worldwide UFO controversy assures sceptics will continue dishing up every possible motive they can to reject the evidence and proof that ET spaceships are here, no matter the proof, or how compelling the evidence. "We cannot know anything because we do not have the original film to analyse," they say. After superficial observation, they claim, "We know he made models because we have made models that look like his photographed objects." There will, however, always be curious people who want to find hidden details available in different copies of the film, and people who want to see the truth, and the truth eventually always wins out. We now share our detailed findings that conclusively show this is a big flying disc performing these movements, not a little model or any known flying device currently available in our civilisation.

The Beamship Demonstration

Introduction

As explained, the Meier case has contributed dramatically to the worldwide UFO controversy, indeed more so than any other UFO case. *The Pendulum UFO* film is a crucial component of this controversy.

On 18 March 1975 on a cold and hazy, or snowy evening, Billy Meier says his Plejaren contact Semjase made a demonstration aboard her beamship. Meier, as instructed, placed his film camera on a tripod in a place not far from Hinwil, in the middle of an open field in full view of a couple of houses with a big tree around 20 to 25 metres tall close to them. As explained to Meier, this beamship, and probably also the WCUFO, had a cloaking device creating an invisible field around it blocking light going out of or reflecting from it, making it invisible to all except for a little window of view towards the camera. Other people around could not see the demonstrations, only Meier and his camera. Many people perhaps heard a strange sound, maybe some saw the tree moving, but no one saw the ship, only Meier and his camera. The beamship was most likely previously programmed to perform specific movements mimicking pendulum movements, but with variations, like jumps, pulses of energy, wobbling, and flipping effects.

The beamship was dancing around the tree, and in one moment made a right turn which moved the treetop without touching it. After three minutes and 20 seconds of continuous demonstration, the ship went away, or the film stopped. A few days later, the tree disappeared.

The similarity of movement in this demonstration to a little model hung from a cord has been part of a several-decades-long controversy. Why did the Plejaren perform this demonstration? Was it to show real evidence of their presence on Earth? If so, why mimic a pendulum and why make it invisible to anyone else present at the time? Some people think these ETs make demonstrations to prove their existence to us. We have found they purposely performed this pendulum-type demonstration to create a controversy. And as explained in the introduction of this book, it was part of a transformation process involving us all, to allow us to gradually realise for ourselves that we are not alone in the universe, and to tell us: *They are Here*. In the 1970s, perhaps we were not prepared to understand this; now, in 2020, we are.

How can we be sure *The Pendulum UFO* was a real spaceship and not a trick executed by Meier with the help of a few partners? The following section describes the results of this our second investigation and presents evidence showing it was real.

The demonstration sequence

In this investigation we refer to the sequence in the FIGU documentary full version of the film available here:

https://youtu.be/Gzr3BRUhfy8

This video has a higher resolution than the FIGU demonstration at the FIGU YouTube channel and comes from the same version of the original film.

Figure 90 shows the timeline of the different demonstration events, again indicating the time as minutes, seconds, and frames (MM-SS-FF). Frames go from 0 to 24 (25 frames on a PAL format video). For example, the beamship "flip" occurs at two minutes, 45 seconds, and in the 24th frame.

In the following sections, we refer to these events and compare them with the contents in the other film copies.

The Beamship Demonstration

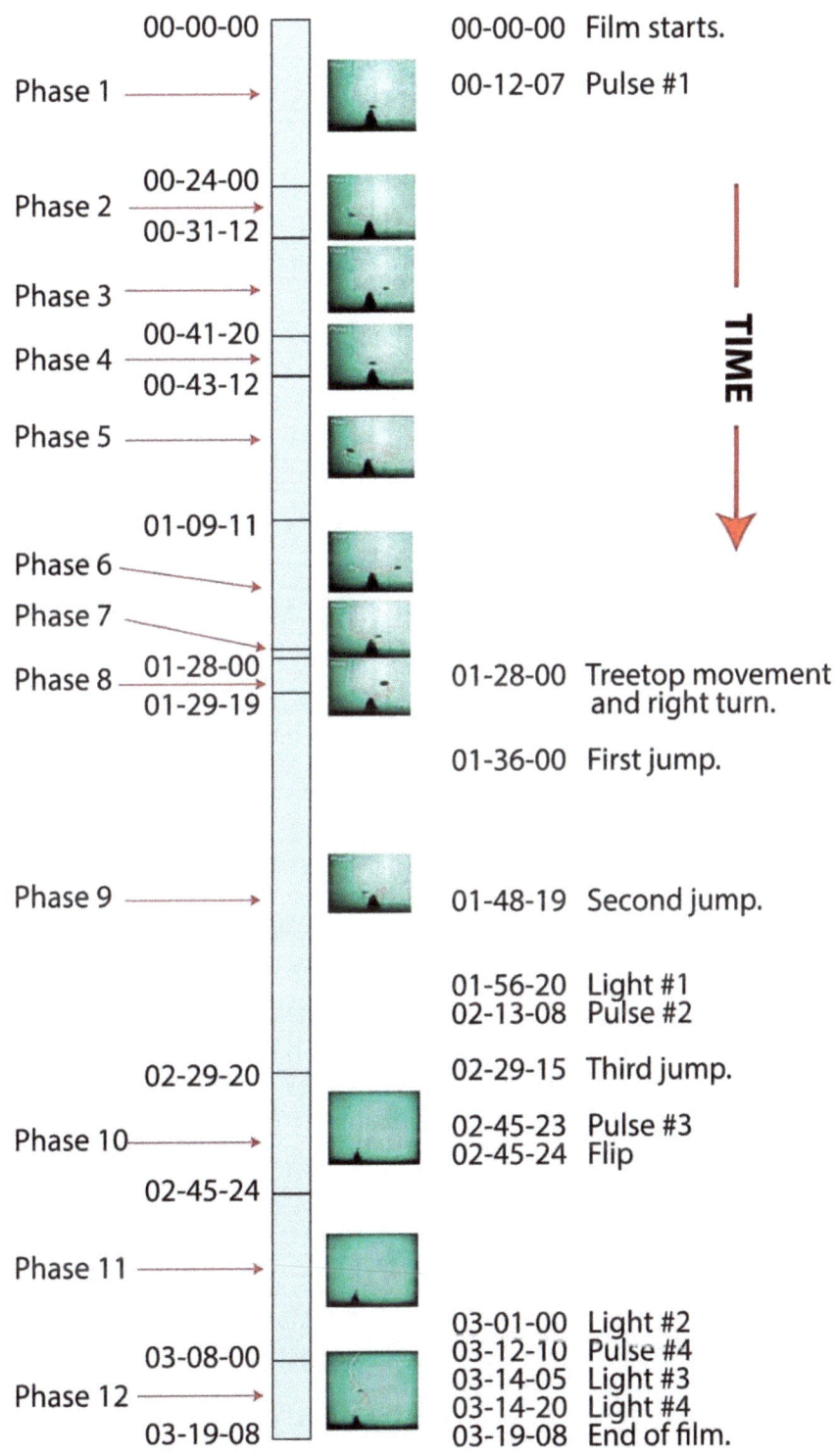

Figure 90 -   Sequence of demonstration events.

Intriguing details

There are eight intriguing details in *The Pendulum UFO* film that cannot be explained by merely claiming Meier used a scale model hung on a cord. They point to the inevitable conclusion that what we see in the film is a demonstration of a sophisticated large flying device. In this second investigation, we now cover these eight points in detail. They are:

1. **Forced pendulum:** The UFO movement is similar to a pendulum, but is not a pendulum.
2. **A big tree and UFO:** The tree, house, and UFO are close to each other, at around 385 metres from the camera. The flying object is not a scale model.
3. **A smooth sharp turn:** The UFO makes a sharp turn while shaking the treetop.
4. **The "jumps":** The UFO conducts three "jumps" close to the treetop. In the 3rd "jump" it stops and remains stationary for 16 seconds.
5. **The sudden UFO flip:** While stationary, it flips five degrees in less than 1/24 of a second.
6. **UFO Wobbling:** The UFO wobbles as if supported from a point near its base, not from a cord above.
7. **The "pulses":** The UFO seems to emit four pulses that create a wide bright arc in only one frame (1/24 of a second). These arcs always cross the UFO image.
8. **Electrostatic flashes:** The camera has or receives electrostatic charges that illuminate the film a few times.

1- Forced pendulum

The movement of this object is similar to but different from that of a pendulum. It does not follow the laws of pendulum physics. Taking into account physical aspects of the object's movements, we now perform tests with a scale model pendulum in various configurations to check the feasibility of using a scale model to create what the video shows. We present video analysis results made with readily-available simple computer tools and discuss a few aspects of the footage that demonstrate the simple model proposition cannot explain what this film shows.

We cover some findings of American physicist Bruce Maccabee and tests conducted by Phil Langdon, two sceptics of the Billy Meier case. We present a sophisticated model that better explains what we see in the Meier film. Our findings lead to the conclusion that the model hypothesis does not and cannot explain the flying object in the video.

Several sceptics, besides Maccabee and Langdon, have tried to reproduce the flight pattern, or provide some explanations, but in their attempts, they have ignored some of the film's essential details. Also, we find in the Meier case that something appearing easy to explain on the first impression is not simple at all upon conducting a detailed analysis. As with the WCUFO analysis, clues reveal *The Pendulum UFO* case is not as simple as one might at first think. The evidence here, as in other pieces of Meier's evidence, on any simplistic initial assessment, leaves room for doubt. After looking in more detail, however, hidden clues arise indicating this is no trick, but rather a sophisticated effort at creating a controversy surrounding UFOs.

We divide the dancing beamship activities into 12 phases explained below and illustrate their time sequence on the left of Figure 90 and in Table 5.

The object moves like a pendulum of the planar, circular/conical and spiral type (see Figure 91), but its period of movement continuously changes. The pendulum *period* is the time required to do one cycle, for example, swinging from left to right and returning to the left, or the completion of a circle. These changes in the *period* in Meier's video are significant; they indicate that the length of the pendulum is continually changing. Table 5 illustrates the twelve phases and pendulum variations.

In our investigation, we measured most oscillations during the different phases in the videos available and performed tests with varying models of scale hung from a cord, which is what sceptics suggest Meier used to perform the UFO demonstration.

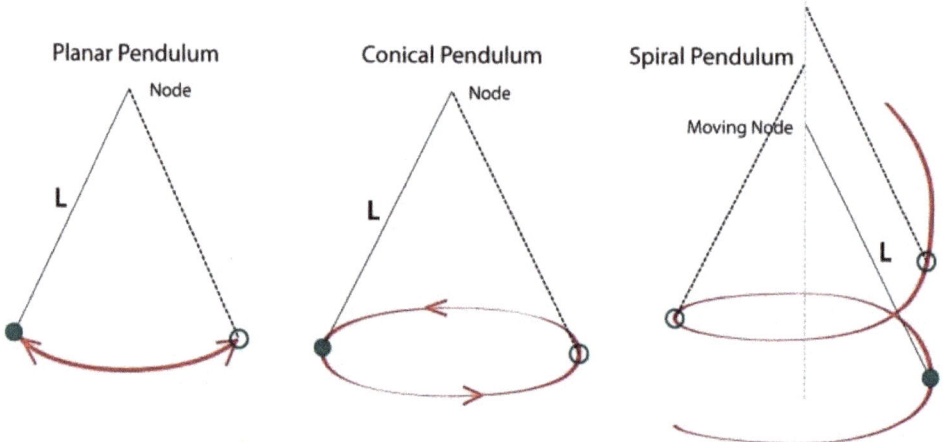

Figure 91 - Types of pendulum movements found in Meier's film. The period in each one follows the same physical law dictated by the pendulum length and Earth's gravity.

Planar pendulum: the model moves from left to right and back to the left in a vertical plane — the pendulum movement of old pendulum clocks.

Conical pendulum: The model moves in a circle, and the cord describes a cone figure.

Spiral pendulum: like the conical pendulum, but the nodal point of cord support moves upwards at a uniform speed.

The Beamship Demonstration

Phase	Description	Image	Pendulum Period (seconds)	Pendulum Length (m/ft)
1	The UFO is stationary for 24 seconds.			
2	It moves to the left as if somebody is pulling it.			
3	It moves, to the right and left, twice, like a free *planar pendulum*.		5.6 5.2	7.8/25.5 6.7/22.0
4	It reduces speed quickly and stays above the tree. It looks like a pull from the right stops it.			
5	Initially, it moves in counter-clockwise circles around the treetop at a short distance. It then increases its orbit diameter in a *Conical pendulum* movement.		4.9 5.0 4.4 5.1 4.3	6.0/19.6 6.2/20.4 4.8/15.8 6.5/21.3 4.6/15.1

They are Here

Phase	Description	Image	Pendulum Period (seconds)	Pendulum Length (m/ft)
6	It gradually switches to a planar pendulum movement, left to right, similar to Phase 3.		4.5 5.0 5.2	5.0/16.5 6.2/20.4 6.7/22.0
7	The UFO moves the treetop without touching it.			
8	Immediately after moving the treetop, it quickly changes direction to a perpendicular plane, moving back and forth instead of left to right. It looks as if somebody pulls it from behind the camera.			
9	It continues moving back and forth and performs three separate "jumps" when passing above the tree. It looks like three cuts in the roll of film. It stops in the final jump.		5.0 5.2 5.3 4.9 5.4 5.5	6.2/20.4 6.7/22.0 7.0/23.0 6.0/19.6 7.2/23.8 7.5/24.7

Phase	Description	Image	Pendulum Period (seconds)	Pendulum Length (m/ft)
10	The UFO remains stationary above the treetop for 16 seconds.			
11	It suddenly twists around five degrees and restarts movement towards the back, again moving back and forth.		5.5 5.5 5.6 5.4	7.5/24.7 7.5/24.7 7.8/25.5 7.2/23.8
12	It departs, moving in a *spiral pendulum* path while ascending.		5.5 5.2	7.5/24.7 6.7/22.0

Table 5 –   Billy Meier's film phases with pendulum periods and measurements.

Measuring the period of oscillation

To accurately measure these periods, we used the video editing tool "Pinnacle Studio, Ultimate Collection", but any readily available similar tool, for example, Adobe Premier gives the same results. It helps confirm the video details if the tool can do a frame by frame review.

They are Here

We defined control points by looking at each frame to determine the exact moment, or frame, when the UFO was at the beginning of the cycle (the leftmost or rightmost position), and then the whole movement until the UFO comes back to the initial position. Our analysis calculated the error in measuring these periods was around 0.1 of a second. In the worst case, where it was difficult to determine the control point, the error was 0.2 of a second.

A relationship exists between the Earth's gravity and the pendulum period and its length. A simple formula to calculate the pendulum length based on the *period* is readily available in physics books, but for ease, this simple formula is available:

$$L = 0.2482 \, T^2$$

Where **L** is the pendulum length in metres (the length of cord), and **T** is the pendulum's oscillation period measured in seconds.

For example, if **T** is 5 seconds, this formula gives 6.2 metres length. With an error of 0.1 seconds in the pendulum measurement, the error in length calculation is 25 centimetres. However, there are variations up to 3.2 metres or more in Meier's film, which means 13 times the accepted error. So these variations are real and significant.

Pendulum oscillations always follow this relationship because it relates to the Earth's gravitational force; the preciseness of movement is why antique clocks incorporated pendulums.

It is essential to keep in mind here that each *period* measured in seconds corresponds to a unique pendulum length value in metres, as indicated in the above formula and the results of Table 5. So, with an object hung from a cord, we can know the cord's length by measuring the *period*, the time to complete one full cycle.

So, the values in the table are accurate and show noticeable variations making the essential hypothetical "model" pendulum movements impossible variations of the pendulum length. The longest distance was 7.8 metres (25.5 feet), and the shortest 4.6 metres (15.1 feet). A difference in the pendulum length of more than three metres (10 feet) occurs. This significant value creates a singular difficulty for the "model" theory.

Three pendulum models:

Three possible pendulum simulations of what Meier's film shows are:

- The fishing rod model.
- The model fixed from a tree branch above.
- The sophisticated model.

The *fishing rod model* consists of hanging a scale model UFO of around 30 centimetres in diameter from a thin cord attached to a long flexible bar, like a fishing rod. Moving the pole accomplishes the different movements. Pulling the cord or elevating the fishing rod at the end of the demonstration moves the UFO upwards, hopefully, as shown in Meier's film.

The *model fixed from a tree branch above* suggests the use of the same or a similar scale model hung from a cord passed over a tree branch outside the camera's field of view. Another cord is attached to a model extremity, allowing an operator to pull it to oscillate the model in the three different pendulum motions of planar, conical and spiral. The cord needs pulling when the filming session ends to correctly catch how the model ascends at the end of Meier's film.

We think our *sophisticated model* is the most practical possibility. It requires a very complicated arrangement with several assistants as in a Hollywood recording studio. We explain this model in detail momentarily.

The fishing rod model

A first impression upon watching *The Pendulum UFO* movie may be that it is a model hung from a fishing rod type of arrangement. Figure 92 illustrates three variations of this model.

They are Here

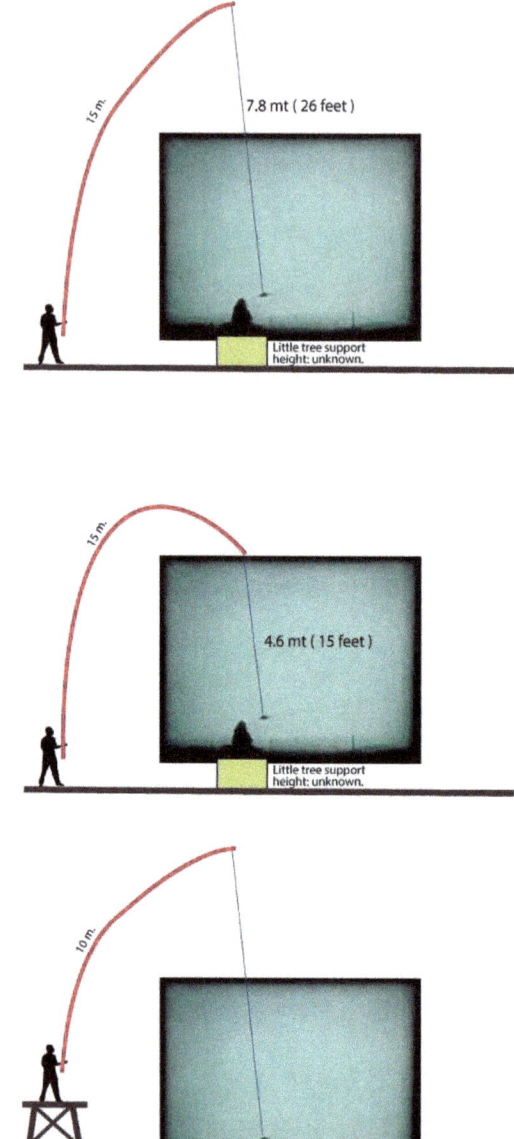

Figure 92 - Variations of *the fishing rod model*.

Maccabee, in his paper "Pendulum-like Motion of an Unidentified Object (UO) Filmed by Billy Meier," points out the differences of the pendulum *periods* in his investigation, which, upon reading, gives the impression that he assumes Meier used a pole, like a fishing rod, from which he hung the model. The variations in time cycles, he explains, are due to the continually moving supporting point or pendulum node, or by pulling a cord. The *fishing rod* method, however, was found not to explain every movement seen in Meier's video. Our study found differences in some time values indicated by Maccabee. We assume he did not analyse with a computer tool as we do in this investigation but instead referred to some *period* variations made from looking at the Japanese investigators' time counter in the Yaoi documentary.

Noticeably, in Phase 12, when the beamship ascends in a *spiral pendulum* movement, the pendulum period does not reduce. In this final phase, if the fishing rod operator pulls the cord to raise the model, the cord length reduces, thus shortening the pendulum period. In Meier's film, however, it remains the same and is even longer than in other phases. So the fishing rod operator must move the rod or pole up without pulling the cord, and this creates a difficulty because the pole must be at least 15 metres long (Figure 92, top). Figure 95 shows a necessary box on a table for a little tree because, viewed from the camera, the ground near his camera is lower than the house in the background. The extra height of the little tree exacerbates the problem of the fishing rod length.

Also, in Phases 5 and 6 of the film (Table 5), the cord must be around 4.6 metres or less, which presents another complication: the operator must lower the rod or pole without it entering the camera's field of view (Figure 92, middle).

Handling a 15-metre-long rod is extraordinarily difficult. Presently, the most extended graphite carbon fishing rods reach this length, but these were not available to Meier in the mid-1970s (Allfishingbuy.com). Another option might be to use a 10-metre-long pole with the operator on a platform elevated five metres from the ground (see bottom Figure 92). But even a 10-metre-long pole is very complicated to handle, and Meier must construct and erect on site the five-metre-high stand without anyone ever knowing or seeing it. There is, of course, zero evidence of any such endeavour on Meier's part.

Maccabee did not consider the variations in length or the significant variations in the pendulum movements. As far as we know, using a fishing rod model, he also lacked a troublesome practical test to confirm the pendulum periods changed in a similar magnitude to those shown in his film. Notably, each circular movement of the node around the tree in the conical pendulum should not change the *period* at all. However, it changes dramatically in the real film, resulting in a significant change of 1.6 metres in the pendulum length. The UFO also looks to be pulled from the left in Phase 2, from the right in Phase 4, from the front camera location in Phase 8 when it does a right turn and moves the treetop without touching it, and from the back after the *flip*, in Phase 11 when it restarts the movement back and forth. So it seems there must be four additional cables attached to any scale model to accomplish this. Notably, Maccabee analysed the Nippon TV recording that does not show all phases of the film.

With some assistants, we attempted a practical experiment to test the *fishing rod* model by using a pendulum to replicate every phase shown in Meier's video. The phases are listed and illustrated in Table 5. The detailed tests and additional tests and findings are shown in the video at this link:

www.youtube.com/watch?v=IKeutVKFbG0

A pot lid served as a surrogate UFO model. The model chosen was 25 cm rather than 30 cm, but this minor difference is inconsequential. The experiment performed used the pole or "fishing rod" approach (tests 1a and 1b). The lid was hung from a thin nylon cord and then hung from a tree branch (test 2). In tests 1a and 1b (Table 6), the pendulum length was two metres (6.6 ft.), and when hung from the branch, it was 3.8 metres (12.5 ft.). These tests were to check the feasibility proposed by sceptics of a solitary person reproducing all the phases while simulating Meier's video. We did not attempt an exact replication of everything Meier's film shows because it would have been far too complicated to perform.

See the results in Table 6 and the recorded *period* measurements. Videos were made and results reviewed with the same video editing software and methods used with Meier's video.

In test 1a and 1b (Table 6), the model was easy to move. Interestingly, regardless of how we moved the node (the pole's top where the cord is attached), the *periods* were the same in all cycles. The time measuring error is the same as that found in Meier's video, around 0.1 of a second, and no more than 0.2 of a second. We concluded that with this method if the cord length does not change, the *period* changes no more than 0.1 of a second. A 15-metre (49 ft) pole test (Figure 92, top), which would be necessary if the sceptics were right, was avoided because finding and manoeuvring such an enormous pole or rod proved a futile exercise.

The usual present-day very long fishing rod is "only" six metres long when fully extended (Hey Skipper). We say that in this *fishing rod* model, Meier must use a rod or pole at least 10 metres long, which is absurd to try. As mentioned above, there are special graphite carbon rods 14 m ~ 20 m long now but, of course, these were not available in the 1970s; and they cost $200 ~ $300 each, way beyond the meagre means of the financially struggling Meier at that time (Allfishingbuy.com).

We conclude the *fishing rod* model cannot explain what we see in *The Pendulum UFO* film.

Test	Description	Image	Pendulum Period (seconds)	Pendulum Length (m/ft)
1a	Using a pole, with a pot lid as a model. Planar pendulum movement in a left-right direction.		2.9	2.1 /6.9
			2.9	2.1 /6.9
			2.8	1.9 /6.2
			2.9	2.1 /6.9
			3.0	2.2/7.2
1b	The same pole as 1a, but with a conical pendulum movement in a circular direction.		2.9	2.1 /6.9
			2.9	2.1 /6.9
			3.0	2.2/7.2
			2.9	2.1 /6.9

Table 6 – Tests 1a and 1b with the "fishing rod".

The model fixed from a tree branch above

In this scenario, the scale model hangs from a cord without a pole supporting it, and the cord hangs from a branch of a tree, both are outside the camera's frame of view (Figure 93).

Phil Langdon made a gallant simulation by constructing a model hanging from a tree, showing how it can approximate Meier's film. The simulation published on his public YouTube channel shows a model very similar to the one in Meier's film. However, Langdon did not analyse the physical aspects or provide the mathematical analysis which Maccabee did. Also, he ignored a few crucial details, like why Meier's pendulum period continually changes but in his model hanging from a tree branch demonstration it remains about the same at 3.7 seconds, which is not what Meier's film shows.[10]

Figure 93 - *The model fixed from a tree branch above.*

Furthermore, in Langdon's tests, he did partial shoots, not a continuous film running more than three minutes showing all 12 phases, which he would have had tremendous difficulty performing. He also did not show how to move a tiny treetop without the model touching the tree and having the model stay in a smooth movement afterwards. In some of his demos, the model

The Beamship Demonstration

touches the tree and shakes continuously. The right turn his UFO model executes results in the model shaking. In contrast, Meier's film shows a smooth movement which is extraordinarily challenging to replicate, so demanding that no one has successfully replicated it to date. If it proves anything, Langdon's bold attempt proves to the observant eye that Meier did not use this *model fixed from a tree branch above* to shoot his film.

Finally, it is essential to note that there were no useful trees available in the location where Meier shot his film. Still, we wish Langdon well for any possible future tests he attempts.

Some significant difficulties with this model and why it cannot explain what we see in Meier's film are:

- The node is in a fixed position on the tree branch above. In Meier's film, the node, where the cord is attached to the branch or fishing rod, is always moving. See the composition of different UFO positions in Figure 94. The cord direction projects perpendicularly to the UFO's longer axis of its major ellipse. There is no unique location for the node, and some cord directions even project divergently and never intersect. With a cord and node, the cord is continually moving.

- The pendulum periods are changing throughout the demonstration, indicating the pendulum length is changing. However, in the *model fixed from a tree branch above* the length is constant.

- In the final Phase 12, when the UFO model must ascend, the cord needs pulling, which reduces its length, so the pendulum period must decrease. Meier's film does not show this: the pendulum period remains the same. So with this model scenario, rather than pulling the cord, the tree branch must rise, which is impossible.

- As we pointed out in the fishing rod model, there must be four different cords to stop or move the model from the four directions of left, right, front and back. With additional operators using these cords, however, we found the cords quickly become entangled in the little tree.

Figure 94 - Composition of several UFO positions and alleged cords projected from each UFO. None of these lines converges into a single node.

Figure 95 - An arrangement simulated from actual photos for a *model fixed from a tree branch above*. The model failed to produce the variable pendulum length found in all phases.

We also simulated this model (Test 2). Watch the results in the *Dancing UFO – Detailed Investigation* YouTube video (Zahi "OVNI Danzante").

In our second test, we used a flexible tree branch capable of producing model wobbling as in Meier's video. The wobbling did not occur naturally in the test, however, proving that someone must pull the branch down to create this effect while someone else creates the other movements. Using a form of support to replace the soft branch, someone must pull and release the cord where the model hangs. Table 7 shows that the *period* duration does not change in most cases. It changes at the end, in Phase 12, since pulling the nylon cord reduces the pendulum length making the *period* smaller as the model ascends. In Phase 12 of Meier's film, this does not happen; the period remains approximately the same or only slightly increases.

They are Here

Phases in Test 2	Description	Image	Pendulum Period (seconds)	Pendulum Length (m/ft)
1	The model is stationary for 23 seconds. A fixed wooden post was used instead of a small tree since the nylon cords entangled in a tree.			
2	Pulling the model slowly from the left.			
3	The operator releases the cord and dashes to the other side, behind the camera. The lid swings twice, like a free planar pendulum.		3.9 4.0	3.8/12.4 4.0/13.0
4	The model quickly decelerates and remains above the tree. Pulling the right cord from the right stops it.			

The Beamship Demonstration

Phases in Test 2	Description	Image	Pendulum Period (seconds)	Pendulum Length (m/ft)
5	The craft moves in counter-clockwise circles, viewed from above. It is difficult to move the model this way with only one operator pulling it.		3.9 3.9 3.9 3.8	3.8/12.4 3.8/12.4 3.8/12.4 3.6/11.8
6	The craft/model gradually switches to a planar pendulum movement, left to right, similar to Phase 3.		3.8 3.9 3.8	3.6/11.8 3.8/12.4 3.6/11.8
7	The craft moves the treetop without touching it.	In test 1a, **not simulated here**, we tried to move a bush without success.		
8	Immediately after the craft moves the treetop, it quickly changes direction to perpendicular. This part is very challenging to replicate. If pulled, the model oscillates violently.			

Phases in Test 2	Description	Image	Pendulum Period (seconds)	Pendulum Length (m/ft)
9	While continually moving back and forth, the craft "jumps" three times when passing above the tree. Only by cutting the roll of film is jumping simulation possible.		3.8 3.9 3.7 3.7	3.6/11.8 3.8/12.4 3.4/11.1 3.4/11.1
10	The craft remains stationary above the treetop for 16 seconds.	**Not simulated here.**		
11	The craft suddenly twists around five degrees. Then restarts movement towards the back, moving back and forth again.	**Not simulated here.**		
12	The model departs, moving in a spiral pendulum path while ascending. The *period* decreases with a decrease in pendulum length, unlike in Meier's film where the *period* remains constant.		3.8 3.4 3.0	3.6/11.8 2.9/9.4 2.2/7.3

Table 7 – Test 2 Phase simulations 1 ~ 12 with a nylon cord hung from a flexible tree branch.

The Beamship Demonstration

When we performed this test, the full-length video was unavailable to us, so the 3rd jump, the UFO's sudden stop, the *twist,* and the *pulses* were not observed, so we did not attempt to simulate them.

In conclusion, neither the Langdon nor the Maccabee investigations offered good explanations of how the model hypothesis could be consistent with the variations in the object's movements – its different *periods*. They also, as previously stated, do not demonstrate the "model" moving the treetop without touching it, and they fail to explain other remarkable details found in this second investigation.

The sophisticated model

We created a *sophisticated model* demonstrating how to include various vital aspects of the UFO's pendulum movement. The approach is theoretical, challenging to accomplish, costly, time-consuming to construct, and is ultimately unlikely to produce results that precisely match what Meier's film shows.

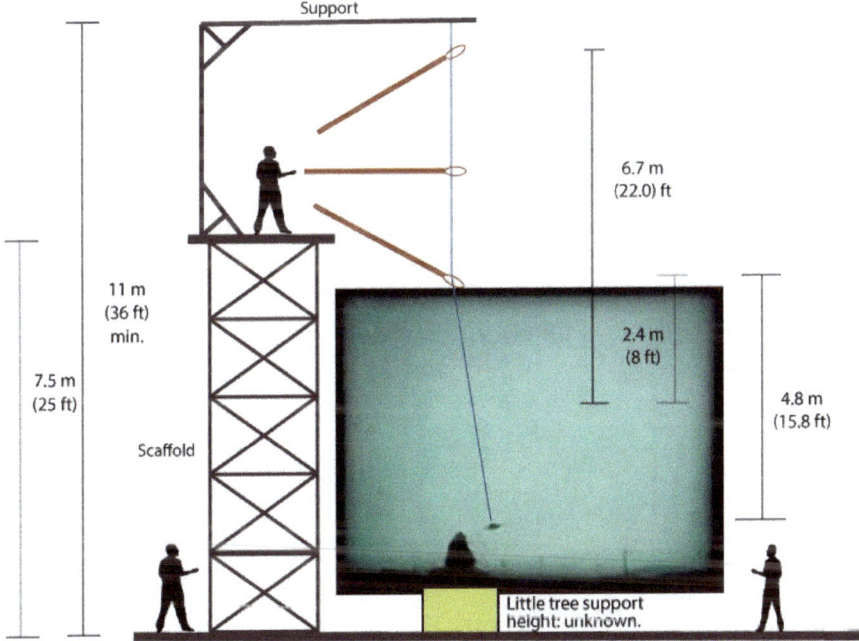

Figure 96 -   A scaled, *sophisticated model* arrangement to make *The Pendulum UFO* video.

They are Here

The model has overhead support for the pendulum cord that is controlled by an operator standing on top of a high platform. This *sophisticated model* requires several operators, a scaffold, and several different cords, as shown in Figure 96. The primary mechanism consists of a long horizontal bar with a ring at one extreme (Figure 97). An operator standing on an 11-metre-high scaffold platform controls most of the model UFO's movements with the bar. As the bar moves up or down, the ring, being the pendulum cord node reduces or increases the pendulum length. Such an operation could demonstrate the various pendulum periods and how the node changes its position (Figure 94).

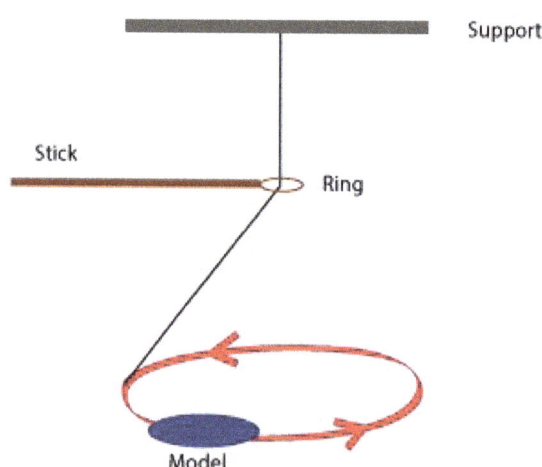

Figure 97 - The overhead support in the *sophisticated model*.

So, could Meier have operated this with just one helper, one of them on the ground and the other on the platform? Unfortunately, no: more participants are required. In Phase 4, the model abruptly stops as if someone pulls it from the right. Langdon's simulation stopped the model in this phase, from the left, not the right. So the operator on the ground needs someone helping him from the other side. Or he could run left to right behind the camera in 10 seconds to hold another cord there, at the right. In this *sophisticated model,* he is the *runner*. So, the model could have just a platform operator and one *runner*.

Our experiment required three nylon cords, one from the top branch to the model, one from the model to the left, and one from the model to the right. It was very challenging to avoid entangling all the cords on the miniature tree below the moving model. It was so problematic that in our test, we switched to a vertical wooden post instead of the tree. Our field test simulated the phases found in the available evidence. With a full-length version of the film, we notice another necessary cord to the back, and another operator there possibly lying down to avoid being filmed. This operator pulls the model from the back, as seen in Phase 11. Five cords are needed: towards the top, the left, the right, the front, and the back. There is a platform operator, and four operators on the ground, ensuring the cords do not loosen and become entangled in the little tree. All must coordinate precisely. If one of their cords is too tight, it stops the model, and if too loose, it intertwines in the little tree, ruining the recording session.

Now, in Phase 7, the UFO passes above the tree and the treetop moves without the UFO touching the tree. Why does the treetop move? Does the UFO model hit a transparent and invisible extended pole from the little treetop? If so, the model would wobble after the hit, but the tree moves and the UFO does not wobble then in Meier's video. Maccabee acknowledges this part of Meier's video but omits explaining how it happens. Langdon's simulation shows the model *hitting* the treetop, which does not occur in the original video. We observe the treetop move after the model or UFO has passed it by one UFO or model diameter. The whole tree also shakes, as shown in the *Contact* movie clip. The results of this investigation might suggest the treetop moved due to air turbulence from the closely passing UFO, or an unconventional force, like a UFO force field. The model we tested and any model in the *sophisticated model* scenario, however, is too small to create turbulence, while a more significant object like a 7-metre UFO could.

A debunker might attempt to explain this by claiming: somebody from the right pulls another cord attached to the top of the miniature tree at the precise moment the object passes over it. However, another firm cord would probably then be required, attached to the middle part of the miniature tree from the left to prevent it from toppling when jerking the cord on the right, unless

They are Here

the tree is secured to a massive base. Accepting this possibility, we have someone else involved: a "treetop pulling assistant," who could be one of the four ground operators. So now at least five people participate in a very complicated simulation; however, it is a theoretical possibility.

Now, how can the wobbling be achieved? The model moves up and down while moving left to right. It can be done by another assistant pulling up and down an additional cord attached to the upper support that must be flexible; if it wobbles the model would do the same. Alternatively, maybe any of the ground operators could do it.

Figure 96 shows the reconstruction of a possible arrangement Meier could theoretically have used to record his video. Three operators are present, but five are needed. From Table 5, we have the minimum pendulum length of 4.8 m (15.8 ft.) in phase 5. This distance helps locate the lowest possible "platform operator" height and the ring at the end of his or her pole. Also, noticeably in phase 12, the pendulum length was 6.7 m (22.0 ft.) while the model was ascending, assuming a representative location midway up the move at 2.4 m (8 ft.) below the lower position of the platform operator's ring. Thus, we find the upper ring position where the platform operator works.

In this arrangement, the platform operator, holding a 4-metre long (13 ft.) pole, must be located at around 7.5 m high (25 ft.), and the top cord support that sustains the model must be at the height of 11 m (25 ft.) Photos of the movie location evince the miniature tree needs elevating from the ground (see Figure 95 arrangement simulated from actual photos). Or the tripod or movie camera needs arranging at a slight elevation. Either way is feasible. The UFO model must be around 30 cm in width (1 ft.), and the miniature tree must be 1.3 m tall (4.3 ft.) at around the same height above ground.

The platform operator performs a few critical roles. They can move the pole in circles, so the model moves in circles too, as in Phase 5, sometimes lowering the pole and sometimes raising it to account for the differences in the pendulum length measured from the model to the ring at the end of the pole. Also, towards the end of the demonstration, in Phase 12, while one of the operators on the ground slowly pulls the cord to raise the model, the operator must elevate the pole ring to prevent the pendulum length

The Beamship Demonstration

reducing. Raising the ring is more easily performed than moving the top support upward mechanically. All this sounds quite complicated, and it is, but in our research, it proved the easiest and most accurate way to match the observable movements of *The Pendulum UFO* aka *Dancing UFO*. Any recreation must be accurate in all details.

This arrangement makes it doable, even if it seems like a Hollywood movie stage set. However, doing it in one single staged shooting seems most unlikely due to the difficulties and complexities of the operation. Anyone would naturally ask, "Why didn't Meier do something simpler?" Is it reasonable to assume Meier constructed this high and bulky platform over only a day or two, with some associates sworn to secrecy and beyond any independent external observation in the middle of a farm, with visitors always passing through and a few nearby buildings with a clear view of his place on a cold and snowy day? Remember, the location of this demo was only 200 metres from Hinwil, so it is not an isolated or inconspicuous area for such a construction. People around Meier always watch him, making it impossible to construct such a set, make the film, and then deconstruct the apparatus with a handful of helpers without anyone seeing the complicated operation.

In summary, this complicated scenario for shooting the video comes from finding the continuously changing pendulum length. Neither the "fishing rod" model nor the "fixed from a tree branch above" model explains these variations in the pendulum length. The fact that at the end of the demonstration (phase 12) when the model ascends, it does not reduce the *period* duration, means the pendulum node (the ring) is also rising in any model scenario.

Moreover, if Billy Meier had employed five assistant hoaxers, then during the 38 years since filming, we would expect at least someone to have blabbed the truth to friends, Meier-case critics, or investigators. Or to have bragged about fooling many Meier-case supporters if not for money from a marketed "expose" interview. So we must assume that any assistant hoaxers remain silent, assuming they ever really existed, which considering the lack of any evidence we consider most unlikely.

A computer animation simulating this proposed *sophisticated model* is available on the detailed YouTube investigation video: *OVNI Danzante – Dancing UFO* (Zahi). At the time of performing this simulation, we did not know of the third "jump", the sudden

They are Here

UFO flip, and the probable pulses the UFO emits four times that leave an arc of light on the film in a fraction of a second.

This *sophisticated model* comes closer to explaining what we see in Meier's film. Other aspects of the demonstration, however, remain impossible even with this expensive, complicated, time-consuming arrangement. We, therefore, must conclude that the pendulum movement we see in Meier's film is not simple, but a forced pendulum movement which cannot be explained by unsubstantiated and untested claims of Meier hanging a little model from a cord.

If you think we might be going a bit far with our *sophisticated model*, we should point out that our findings echo the words of top Hollywood special effects people. Here is a quote (provided by Michael Horn) from Special Effects Academy Award-winners for the movie *Independence Day* Engel and Weigert on *The Pendulum UFO*:

> But, to reflect on the statement that's in the film, I also remember seeing a shot on the Super8 reel that showed a UFO circling around a fairly tall tree." According to that shot, we said that we can't conclusively say whether it's real or not, but it seemed impossible to stage that kind of a shot with a miniature (it would have to be hanging on a very tall crane, with wires - but even then the movements would be hard to achieve.) So, yes, in regards to that shot, we mentioned that we could definitely do it today with CG, but at the time these were supposedly shot - it would have been very hard, probably even impossible, to fake this kind of shot (Engel and Weigert Uncharted Territory 2006).

We can also add to these comments that using a big scale model, of seven metres, pulled from a very tall crane will show a pendulum movement of some 20 to 28 seconds, not 5 to 6 seconds. In other words, the motion of a giant model would be prolonged, and this is not what we watch in Meier's film.

2- A big tree and UFO

After watching every version of the available videos, we note the tree, house, and UFO are far away. The UFO moves the treetop, demonstrating an interaction between the two and confirming their proximity in the distance. With the tree more than 20 metres tall, by comparison, the UFO is around seven metres in diameter.

Atmosphere attenuation in the form of Rayleigh light scattering causes blurriness, softer tones, and loss of vivid colour in more distant objects. Light scattering, discussed in more detail below, confirms the tree, house, and UFO are far from the camera.

A big tree close to the house:

Figure 98 - An enhanced image shows Rayleigh and Mie light scattering effects. **A**: The background sky. **B**: The distant house, tree, and beamship in light blue-grey tones. **C**: Image enhancement reveals more vivid colours hidden in the film in the closer foreground than in A or B.

They are Here

Close observation of the film's better videos soon confirms the tree's reality. Figure 98 is a computer-enhanced zoomed screenshot image showing Rayleigh and Mie light scattering effects. We see a transparent background sky **A** with faint and fuzzy very distant towers at the bottom right. It shows the house, tree, and UFO **B** sharing the same degree of blurriness and light and blue-grey tone density, making them about equidistant from the camera. The light rain or snow at this time adds Mie light scatter to the Rayleigh atmospheric attenuation. These effects largely account for the noted blurriness and light blue-grey tone density of the distant objects. Note the different tonal qualities in this film. Compared with the lighter-toned horizon, the UFO, tree, and the house have approximately the same dark tonal quality. An object much closer, like the surrounding grass **C**, or a model tree or model UFO, will be even darker, or more vivid in colour for a colour film. The enhancement in Figure 98, with no colour input or change, reveals, hidden in the foreground grass of the film, more vivid colours than in **A** or **B**, demonstrating the greater distance of the UFO, tree, and house.

Figure 99 - Left: A simulation illustrating how a little tree and scale model close to the camera would look. Right: Actual image from Meier's film enhanced by increasing contrast and brightness and no colour input or changes.

The Beamship Demonstration

These are facts dictated by the nature of Rayleigh and Mie light scatter, which explain how clarity, tone density, and hue quality or vibrancy all deteriorate with distance. [11] Although the film is of relatively poor quality, enough difference in tone and colour quality exists and knowing this tells us that the house, the tree, and the UFO are all approximately at the same distance from the camera.

Figure 99, left, shows a simulation of a miniature tree and UFO model close to the camera. Their closeness to the camera makes their images sharper and more vivid in colour and tone, although exaggerated here to illustrate the point. This close, their level of blurriness and tone should not be the same as the background house which is subject to atmospheric attenuation on this dimly lit evening with light snow or rain. Yet in Meier's film (Figure 99, right) the UFO, house, and tree are all approximately at the same degree of blurriness, tone, and colour degradation.

This simple fact confounds the claim that Meier might have purposely focused on the distance to make the nearby tree and UFO blurred. If he had, why is the house so blurred? It should be in focus. The claim is invalid because the house, tree and UFO all share the same qualities. Conversely, if Meier had focused on a foreground miniature tree, the tree would still be much darker and reveal more hidden colour with enhancement, and the house would be more out of focus than the tree, which is not the case.

Furthermore, in the same figure, right side, we see no evidence of any supporting device for the tree like a table or plant pot, and part of the house is visible beneath the tree's lower branches in some versions of the film. Figure 99, right, shows the actual image from Meier's film enhanced only by increasing brightness and contrast, without any colour input or changes. Try it yourself.

Today we can see the houses, the distant towers, and a forest on the left, but the big tree is missing. Figure 100, left, shows an image from the Nippon TV documentary compared with *The Pendulum UFO* film image on the right. Simulations made by sceptics like Langdon did not simulate a distant fuzzy house in the background during a light rainy or snowy afternoon with a little tree, and small UFO model close to the camera with a similar degree of fuzziness and loss of colour and tone due to Rayleigh and Mie light scatter. A demonstration using a miniature tree and model close to the camera on such an evening would produce something more like what is seen in Figure 99, left side, minus the excessive light on the tree.

They are Here

The only conclusion to arrive at here is that this film does not show a miniature tree and model UFO close to the camera and far away from the distant house.

Figure 100 - Left: Image from the Japanese investigators' film showing the distant house sans the tree. Right: Meier's enhanced film image. The UFO, and tree with no supporting device, seem close to or even behind the house, which is partly visible beneath the tree's lower branches.

The moving treetop

In Phase 7, as previously mentioned, the UFO moves the treetop without touching it. The two interact, showing that the tree and the flying object are at the same distance from the camera. Maccabee recognises this but does not further explain the matter. All available videos show the tree shaking after the UFO executes the right turn. The one showing it most clearly is the clip from the documentary *Contact*. We see not only the upper part of the tree moving but also that the whole tree is shaking in the way a big tree does not a miniature or bonsai. See it in this clip:

https://youtu.be/gGCuLIVxxQw

One of Langdon's simulations shows his model hitting the top of his miniature tree to move it, which is not what happens in Meier's film.

We performed tests with a small, flexible bush. As our model moved very close to and above the bush, any slight turbulence created by the model did not move a single tree leaf. A tiny object like a model cannot generate enough turbulence to move it. However, a significant object can more easily produce this turbulence due to the more substantial amount of air it displaces.

Imagine wearing a hat and a remote-controlled scale model aeroplane passes over your head. The hat stays on your head because the turbulence is not strong enough to remove it. Imagine, however, that you are on a landing strip in an airport, and a plane passes over you, or even someone two or three times your height. Not only do you lose your hat, but the turbulence also moves you. Alternatively, you have probably had the harrowing experience of standing by the roadside as an enormous truck roars past and its turbulence blows off your hat or catches inside your umbrella, knocking you off balance. It takes scaled up large objects to create turbulence; a model may move a few specks of light dust, but nothing much of any weight; a small model cannot drag enough air and create wind.

Watching Meier's film in slow motion, it is noticeable that the tree moves after the UFO passed, as the movie frame shows in Figure 101 when the treetop starts its movement. The tree's motion is consistent with turbulence caused by a big object. The UFO shifts the air around that moves the tree. So a delay can be expected in the tree's movement. A possible distance between the tree and the sizeable UFO could account for the time delay as the air travels to the tree to cause its motion.

However, it is not impossible to simulate this effect in a miniature tree. Pulling a wire attached to the top of a tree, as explained in the *sophisticated model* scenario might imitate it. Perhaps it could be done, but it is so complicated that no one has done it to date, and the results would not be expected to be as good as the original film for reasons mentioned earlier. Synchronising this movement at the right time is very difficult. If Meier did it as in the *sophisticated model* theory, he would have lost many rolls of film before catching the model and treetop at the right moment in time, and remember this phase is not the only footage in the film. Of course, the *sophisticated model* theory requires another operator with a cord located at the right of the tree waiting for the precise time to move the little treetop. As we have shown, a one-armed man cannot possibly perform this entire procedure on his own.

Figure 101 - Treetop movement after the UFO passes above it.

Some investigative supporters of the sizeable UFO theory do not fully agree with the "turbulence" explanation. They say that there may be another mechanism from an extra-terrestrial spaceship capable of moving an object without touching it, like, a force field, energy dispersion, or something similar but entirely unfamiliar. It appears that only one of the UFO passes was too close to the treetop – the Phase 7 pass – close enough to produce turbulence to move it, assuming turbulence was the cause.

According to Meier and the UFO hypothesis, the tree was accidentally exposed to some form of radiation when the UFO approached it too closely. Meier says this was the reason for its removal; the radiation gave evidence or proof that the ETs were there, and this went against their plans or directives. Unable to eradicate the radiation, they eliminated the tree to erase their

The Beamship Demonstration

radiation "fingerprint". If this happened, then the UFO indeed interacted with the tree in at least an energetic radioactive manner of some kind which could have produced tangible results such as the tree moving. Keep this in mind, because it could prove significant when later looking at the UFO "jumps" performed, and the burning produced on at least one of the film frames during a UFO "jump". We discuss later these "pulses" that create an arc of light, and the evidence suggesting electrostatic charges inside the camera.

The disappearance of trees in the Meier case is an enduring mystery. Guido Moosbrugger in *The Silent Revolution of Truth*, movie (Horn), describes how he witnessed a demonstration of one big tree that was vanished by Semjase. This activity from the Plejaren, of course, has contributed to the worldwide UFO controversy, giving rise to sceptics claiming Meier used little trees and small models. However, the evidence of the big tree being close to the house is undeniable.

To conclude this section, turbulence from a big object, not a scale model, is the more plausible explanation here. However, since a UFO can fly without disturbing the air, possibly the proposed air turbulence is not what happened, and other unknown UFO forces are at work. When dealing with technology beyond our understanding, there are many possible explanations beyond our ability to describe or understand.

The UFO is not a scale model

As shown earlier, the UFO, the tree and the house are at approximately the same distance from the camera, which can only mean that the UFO is a big object, not a small model close to the camera; however, we now perform a double-check.

Meier reported that it was snowing lightly on the day he made his movie. The video shows that despite a little clarity in the air, some haze caused by light snow or rain results in Mie light scattering, which in addition to the Rayleigh light scattering causes deterioration in visual or image quality due to light scattering off the rain, mist or snow (Wiki). Figure 102 shows a comparison of the distant house and hills on a clear day from a film recorded by Japanese investigators (Yaoi) in the same place Meier recorded his video. The hills at the right are noticeably blurry in Meier's film. The air was less clear due to Mie scattering caused by the inclement weather, which partly explains the considerably blurred images in Meier's film, which he also filmed in the approaching evening as the light begins to fade.

So, it is unlikely, and there is no evidence to suggest that the camera was out of focus on purpose as some would-be debunkers claim. What we do have is low visibility due to an overcast late-day sky, exacerbated by light scatter from inclement weather, plus the distance of the subject from the camera.

Figure 102 - Comparison of the location's background blurriness.

The level of blurriness or sharpness of objects visible in the film hints to whether the objects are distant or nearby. We noted earlier that the UFO, the house and the tree all show a similar level of blurriness and grey tone quality, indicating their relative proximity. Also, we note the UFO changes in clarity due to atmospheric attenuation and light scatter depending on whether it is closer or farther away. The even darker and greener flora at the bottom of the film (Figure 102, top), due to less Rayleigh light scatter, prove to be the closest thing to the camera, making the tree, house and UFO farther away, precluding the possibility of a small model.

Maccabee used a formula to calculate the distance and size of the Unidentified Object (UO) based on his assumption that it is a model on a pendulum. His calculation showed that the object was very close to the camera. He calculated that the distance between the UFO's most distant and nearest points in its circular movement around the tree is ten times the diameter of this circle. He estimated this value based on the apparent difference in UFO/UO size when close to when far away. All these calculations, however, can be equally applied to either a nearby model (UO) or a distant substantially-sized UFO.

The geometry shows both scenarios produce the same results, meaning, the proportions are the same for the UFO and the model when comparing their closest and most distant positions. Figure 103 illustrates both possibilities: a distant UFO, and a nearby model. The small point **O** at the far left represents Meier's camera. Two small circles on its right represent the scale UFO model, with the little tree between them T_1; and the two circles at the right represent the substantially-sized UFO with the large distant tree T_2 between them. The apparent size of the model or UFO when near (left circles) is the angle **P-O-S**, and when far (right circles) is the angle **Q-O-R**. The geometric angles are the same in both cases, making relationships between the apparent sizes when near or far the same for both the small model and a substantially sized UFO.

Maccabee estimated the distance **A1** as 10 feet (three metres), twice the radius of five feet, and the distance to the nearest point 50 feet (15.2 metres). The furthest distance in his calculation would be 50 plus 10, or 60 feet (18.3 metres). He based his calculation on a ratio of 1.2 of the object's apparent size when near compared with when far. Deardorff estimates this ratio to be 1.3.

They are Here

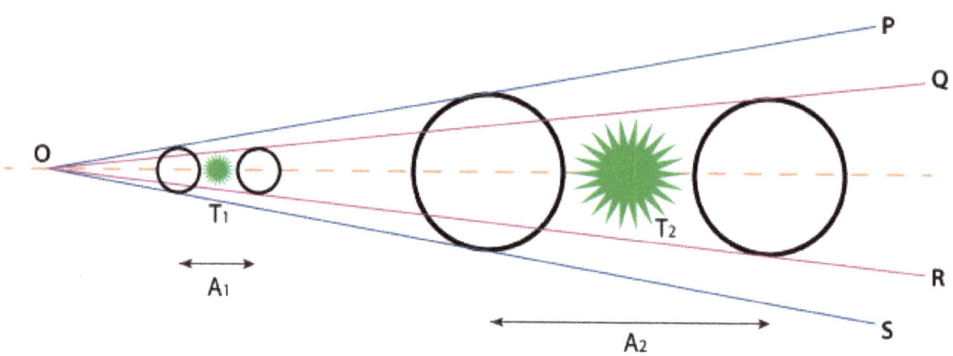

Figure 103 - Left: geometry of a nearby model. Right: geometry of a distant UFO. The angles are identical.

Assuming 1.2 is correct: Doing the same calculation for a distant UFO, we merely have proportionally bigger values. We can estimate the radius of the UFO's circular movement at 35 metres (115 ft.), giving the values in each case shown in Table 8:

Object	Closest Distance	Farthest Distance	Little Tree/ Tree distance
Model (30cm/1 ft)	15.2m (50 ft.)	18.3m (60 ft.)	16.8m (55 ft.)
UFO (7m/23 ft)	350m (1150 ft.)	420m (1380 ft.)	385m (1265 ft.)

Table 8 –   Distances to a model and a UFO.

So:

A1 = 3m (10 ft.)

A2 = 70m (230 ft)

How can we know whether we are looking at a small model or a large UFO when the geometry indicates either could be the right answer? Maccabee uses the pendulum formula from Newton's law. If we assume the model theory correct, and that we see a pendulum movement, then the distances are the ones for the model in Table 8.

On the other hand, if we consider the UFO theory correct with the UFO *simulating* a pendulum movement, the distances are the ones for the UFO in Table 8. Newton's law contributes nothing towards finding the right answer. Either theory still might be correct. However, the level of blurriness and atmospheric attenuation of the images puts an end to this controversy.

Figure 104 (left) shows the images of the UFO when near and far, the house, and the tree (right). The edges of the UFO are noticeably much less sharp when far away (top) than when near (bottom). The difference between the two images is undeniable. The top UFO image is smaller, less distinct, and slightly lighter in tone.

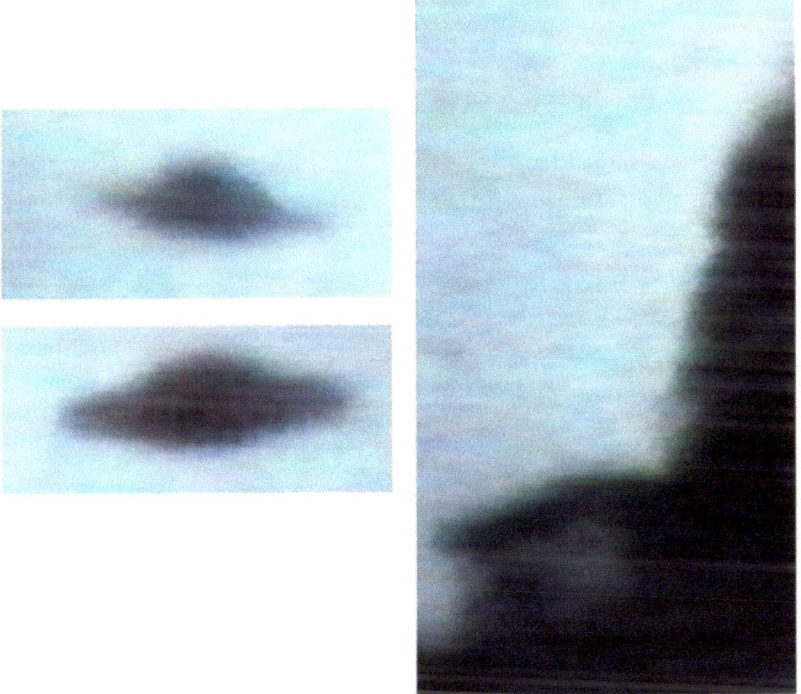

Figure 104 - Different levels of UFO blurriness due to Rayleigh and Mie light scatter. Top left: UFO afar. Bottom left: UFO nearer. Note the differences in size, tone and attenuation. Right: The house, tree, and darker greenery at the bottom.

They are Here

Now, for a model that is not far away from the camera, a difference in the distance **A1** of just three metres (10 feet) does not create such a difference in blurriness or attenuation of an image taken with the movie camera Meier used. The only way that would happen is with very dense fog or smoke in the environment, and if that were the case here, we would not be able to see the distant house or distant hills. See Figure 106 with a simulated foggy environment.

For a significant object of seven metres, however, the difference in distance **A2** now being 70 metres (230 feet), we would, on a dull and slightly snowy evening like this, find a difference in blurriness similar to that in the images; just as Meier's film shows.

The increase in blurriness in the top image compared with the bottom image in Figure 104 is undeniable evidence of atmospheric attenuation, and light scatter. Visually, atmospheric attenuation is the loss of clarity or image degradation that occurs over a distance as light scatters through the atmosphere – Rayleigh light scatter. The more atmosphere the light must travel through, the more attenuation, or loss of image quality, often referred to as turbidity.

The amount of Rayleigh light scatter attenuation at a close-up structure or model is negligible, even from Rie light scatter when the atmosphere is relatively full of particles like rain or fog. It is at a distance that we see atmospheric attenuation take effect. So again, we know that the atmospheric attenuation we see in Figure 104 from the Meier film is not from a close-up model. It has to be from a distance. Here, a range of around 350 ~ 400 metres with a distance of about 70 metres between the two UFOs in the images on an overcast day with slight rain or snow would seem consistent with the degree of atmospheric attenuation we see.

It is perhaps apropos here to diverge for a moment to show Meier's famous never-to-be-reproduced 1976 Hasenbol UFO which shows unequivocal evidence of Mie light scattering. (See Figure 105.) In this photo, we see a deep atmospheric channel from the camera to the UFO with evening mist occupying a fair distance of the air channel and concentrated in visual effect at the atmospheric channel's farther end. The illumination of the distant mist particles over, and in front of, the UFO is typical of a beautiful misty early evening Mie light scatter.

The Beamship Demonstration

Mie light scatter is caused by light passing through larger particles than the elemental oxygen and nitrogen particles of the atmosphere: particles like mist, smoke, dust, rain or snow in the atmosphere present at the time. Here, of course, the sunlight is scattering through the early evening mist.

This mist in Figure 105 is diffuse and light as we see from the distant hills, and it would not show up on a model just one metre in front of the camera. This golden misty light scatter proves this UFO to be some distance away, far enough for the slight mist to attenuate acuity of the image and to scatter the light after passing through a sufficient channel of light mist between the camera and the object.

Figure 105 - Meier's 1976 Hasenbol UFO photo #174, sides and top cropped to show the golden Mie light scatter over its front.

Similarly, the light rain and snow coupled with Rayleigh light scatter cause loss of definition of the images in Figure 104, with the more significant loss occurring in the distant image (top left).

Returning to the movie film and Figures 104 and 106: The top UFO in Figure 104, at around 70 metres farther away than the bottom UFO, is showing the more atmospheric attenuation from Mie light scattering in this slight evening rain or snow. The light rain or snow exacerbates degradation of image quality caused by

Rayleigh light scatter. The degree of attenuation shown in the two images does not occur over a distance of three metres without something very dense in the atmosphere, for example, thick smoke or very dense fog. Figure 104 shows a distant UFO at two significantly different distances. [12]

A simple test can confirm this for yourself. Go outdoors, locate an object at the distances indicated in Table 8, and you find that a nearby object with little variation in its distance does not create such a difference in blurriness or image acuity as a distant object. Significant atmospheric attenuation occurs at times of low visibility or haziness in the atmosphere, as in the heavily overcast evening and turgid atmosphere when Meier filmed his movie.

Figure 106 - The difference in image blurriness between the near and far UFO is only explicable in the model theory with a very dense foggy environment. However, in such a foggy evening, the distant house would be invisible. The little model theory fails to explain the difference in blurriness, and the similarity in the darkish grey tone of the house, tree, and UFO present in Meier's video. A closer model must be darker and more vivid still, like the greenery in front of the tree.

In conclusion, this object can only be substantial in size, and the UFO theory of a large craft is the only hypothesis agreeing with all the available data, so we can only assume it correct.

3- A smooth sharp turn

In phase 8, after the UFO moves the treetop, we see it perform a very sharp turn moving towards the camera (see Figure 101). A suspending cord might be visible using a model, as in the Model Theory, since the line pulling the model is very close to the camera; and we found it impossible to simulate this smooth movement with a model. Furthermore, by watching the Langdon simulation in his YouTube video, we see that the model shakes after this turn, and it looks like the model receives a violent pull. In Meier's film, this movement is characteristically smooth (Figure 107).

Deardorff, on his website, gives more arguments to show that this movement cannot be from a model scenario. It is easy to concur since we tested it several times, and it proved excessively challenging to duplicate and impossible to perform smoothly. The model always tilts and shakes after this turn. We have been unable to reproduce this smooth movement and have never seen anybody else do it successfully.

Figure 107 - A composite of different images of the UFO performing the smooth sharp turn. Image from Deardorff's website.

There is a significant comparison made by Taro in his YouTube channel (Istok 2018; 2019) in which he compares the movement during the right turn shown in Meier´s film with a test using a small model made by Langdon. The model shakes dramatically at the sharp turn.

Some sceptics have suggested that the model could have a gyroscope, eliminating any shake in this sharp turn. It sounds extraordinarily sophisticated for a small 1970s model. Making a convincing small UFO model is complicated but not impossible, but it is highly problematic to include one or two gyroscopes in the 1970s. However, suppose somebody made a small one and installed it into a small model UFO. In this case, the model would be pointing in the same direction most of the time, like its central axis pointing upwards. We do not see this in Meier's film. Most of the time, the central axis points towards a central node, the hypothetical pendulum node, where the theoretical cord is attached to a hypothetical tree branch or fishing pole.

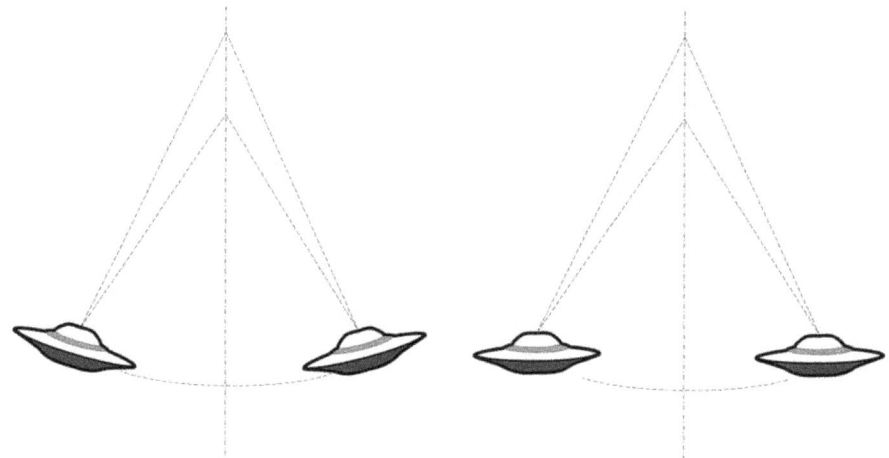

Figure 108 - Left: Simple UFO Model. Right: Complex UFO Model with one or two gyroscopes. Meier's film shows a movement similar to the model on the left.

Figure 108 shows a simple UFO model in a pendulum (left) and a complex UFO model with two internal gyroscopes (right). Most of the time the vertical axis of the complex UFO model would point in the same direction, for example, upwards. Meier's film, however, shows the UFO axis always changing its direction.

4- The "jumps"

UFO "jumps" in space are a fascinating aspect of the demonstration and are evidence of the UFO moving extremely fast from one location to another. This characteristic has also created a big controversy. Alone they do not prove the video UFO is not a small model; sceptics consider the jumps the result of cutting the film roll. Unfortunately, we do not have access to the original film to check for splices at the film's "jump" frames. Even if we found the original film and it had splices, we could not be sure "somebody" had not made splices precisely at the frames where the jumps occur. If we were lucky enough to find the original roll, rather than a copy, and if we found no evidence of a cut in the film, we could conclude the jumps real. On the other hand, even if we found the original movie cut at the "jump" frames, we could not conclude Meier or any specific person cut the film roll because any one of many people could have tampered with the film. It is, however, interesting to check the details of these "jumps" in space.

Three UFO "jumps" occur in phase 9, covering a short distance, above the tree, "jumping" around 10 to 15 metres based on estimates that the UFO measures seven metres. The first "jump" occurs at 1 min 36 sec, the second at 1 min 48 sec, and the third at 2 min 29 sec into the film. The first and second jumps happen in a continuous back and forth movement of the UFO. The third one ends with the UFO stationary above the treetop.

In our first investigation, we found only the first two jumps because our available film copy at the time was not a full-length video. We also found, based on that video that the jumps happened during three to four frames in a gradual transition. With new evidence, however, it seems the jumps happen in just one frame, in less than $1/24^{th}$ of a second, and we now know this video was a 30 fps conversion from a PAL video that produces an interlaced effect described previously in this book.

They are Here

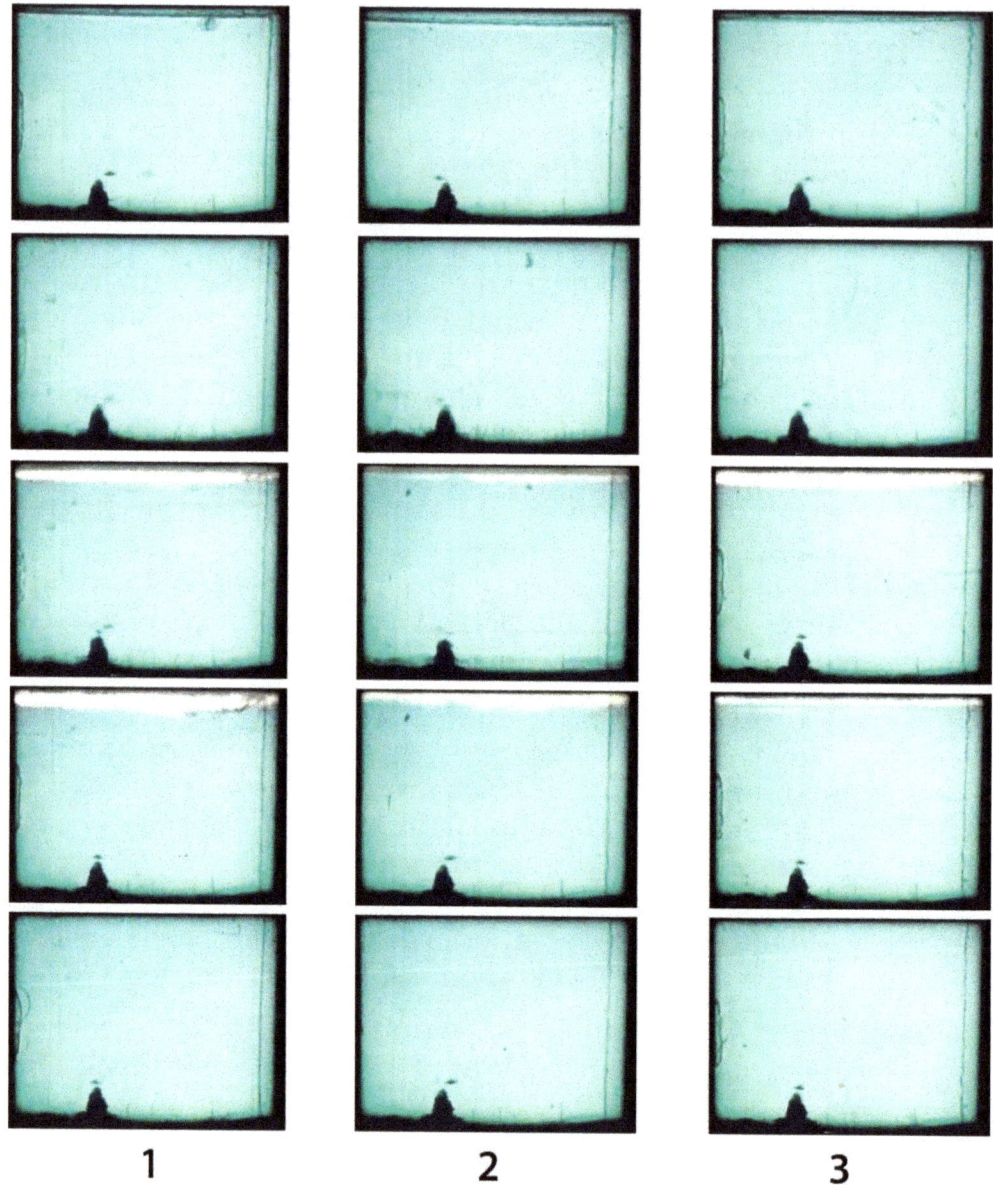

Figure 109 - The three "jump" sequences, from top to bottom, taken from the PAL video *Demonstrationsflüge / Demonstration flights*.

Maccabee considered these "jumps" a trick caused by cutting the film, meaning cutting away a few intermediate frames in the roll and discarding them. He says that the camera jumps in the cut area in the film, and it does. He posits that the cut film was not well aligned, and this is the reason the image jumps. We think

Maccabee is not looking at the whole picture. Analysing these events frame by frame in zoom and full view (the whole picture) evidence reveals this "jump" to be a very unusual event. Also, the full-length documentary shows several splices between scenes with not one jumpy image in them. The camera jumped during the jumps, but the reason why is not apparent.

Figure 109 shows the sequence of the three jumps from the FIGU YouTube channel PAL video. A higher resolution copy is available here: https://youtu.be/Gzr3BRUhfy8.

We find a white band in two consecutive frames (Figure 109) at each jump. In each jump, the original film must have had only one frame with this white band, not two. We see two due to the previously described interlace effect.

Are these white bands the result of splicing the film roll? There is a clear difference between a white band caused by "bad" roll splicing and what we see on the film (compare Figures 109 and Figure 85 left side). Alternatively, could electrostatic charges inside the camera have produced this white band? We found evidence of several flashes from these charges, which we show and discuss in a section ahead.

In the classic documentary *Contact* (Stevens and Elders), when Meier described this beamship jump capability, he says he felt an electric shock when it happened. Maybe these ships produce an electromagnetic pulse (EMP), or some electrical pulse that somehow affects the camera, overexposing or burning the top area of a few frames (the bottom of the camera since the image is inverted). This pulse might also move or alter the normal movement of the roll through an electric shock wave, which could also explain Meier momentarily jolting the camera and the movement of the image. The movie camera has metallic mechanical parts that can be affected by an EMP. An internal electric arc generated inside the bottom of the camera could explain the bright band on the top (the image being inverted).

We now discuss each of the three jumps in detail:

First "jump"

The first "jump" happens in phase 9 when the UFO is moving back and forth at 1 min 36 sec in the film. See zoomed images in Figure 110 of the sequence of four frames from the PAL video (left

They are Here

to right and top to bottom). As the beamship is moving away from the camera, the whole image jumps during the beamship "jump". We see the interlaced bands in this sequence very clearly in Figure 110. The first frame shows a shadowy object about the same size and shape of the UFO to the right of the beamship, which could be just a piece of dust on the film. On the second frame, the beamship has almost disappeared. The interlacing effect could cause this, so ideally more analyses should be performed from the original film, or at least from an excellent Super 8 format copy. The beamship takes less than $1/10^{th}$ or $1/24^{th}$ of a second to move about 10 m between two points.

Figure 110 - The first "jump" sequence in four frames from the PAL video. Noticeably, the whole image jumps in the second and third frames with interlaced lines slightly in evidence.

The Beamship Demonstration

Second "jump"

The second "jump" occurs a few seconds later at 1 min 48 sec. For this jump, the UFO approaches the movie camera and performs a jump similar to the first one. Figure 111 shows the four-frame sequence from the PAL video. The beamship begins behind the tree, and three frames later it is in front, moving 10 metres in 1/10th of a second or less, a speed of around 300 kph or more when it was previously moving and swinging slowly. The interlaced bands (Venetian curtain) are more evident here than in Figure 110, as the whole image and the beamship jumps.

Figure 111 - The second "jump" sequence in four frames from the PAL video. Here the interlaced lines are more evident.

Third "jump"

The third jump is different from the others. The beamship approaches the treetop and "jumps" by disappearing, but upon reappearing ahead, it stops. It is not like a routine vehicle, gradually reducing speed until it comes to a stop; in this case, the beamship stops instantly. Again, sceptics may say it is the result of cutting the film roll. Alternatively, we may say it is the result of unknown technology. This jump happens at 2 min 29 sec in the film. After the jump, the beamship remains stationary for 16 seconds and then performs "the flip" (described later). Figure 112 shows four frames from the PAL video of jump 3.

Figure 112 - The third "jump" sequence in four frames from the PAL video. Here also, the interlaced lines are evident. The object in the last frame above the house at the left is a large piece of dust.

Here again, the whole image jumps and causes the interlaced lines in the third frame. This time the jump distance seems to be shorter, maybe five metres.

How can a beamship perform these "jumps" or stop instantly? We know this is not a small scale model for reasons described earlier, so knowing these jumps are real and not a trick, how are they accomplished? We do not know. Our current knowledge of the laws of physics does not inform us. Meier says that the beamships create their own, and isolated, gravitational field. They function like little self-contained planets with their gravitational field independent of the earth's external gravitational field. This separate gravity field enables them to perform right turns, "jumps" from one location in space to another, or even to stop abruptly without the pilot feeling any effects of inertia. If they jump, not just 10 metres, but significant distances, for example, several light-years, are they violating Einstein laws? Would they be travelling faster than the speed of light? Not really. If they "jump" such distances, they are not strictly speaking travelling. They would be jumping to a different location in space-time, primarily without travelling through space-time to get there, a process we have no understanding of yet.

They are Here

5- The sudden UFO flip

After the third "jump" followed by 16 seconds of remaining stationary, the beamship flips five degrees counter-clockwise in less than 1/24th of a second. It does so precisely when a "pulse" occurs in the film. We discuss in detail momentarily the four "pulses" found in the film. The manifestation of each pulse appears in a single frame as a wide bright arc of light like a flashing line that always crosses the beamship image.

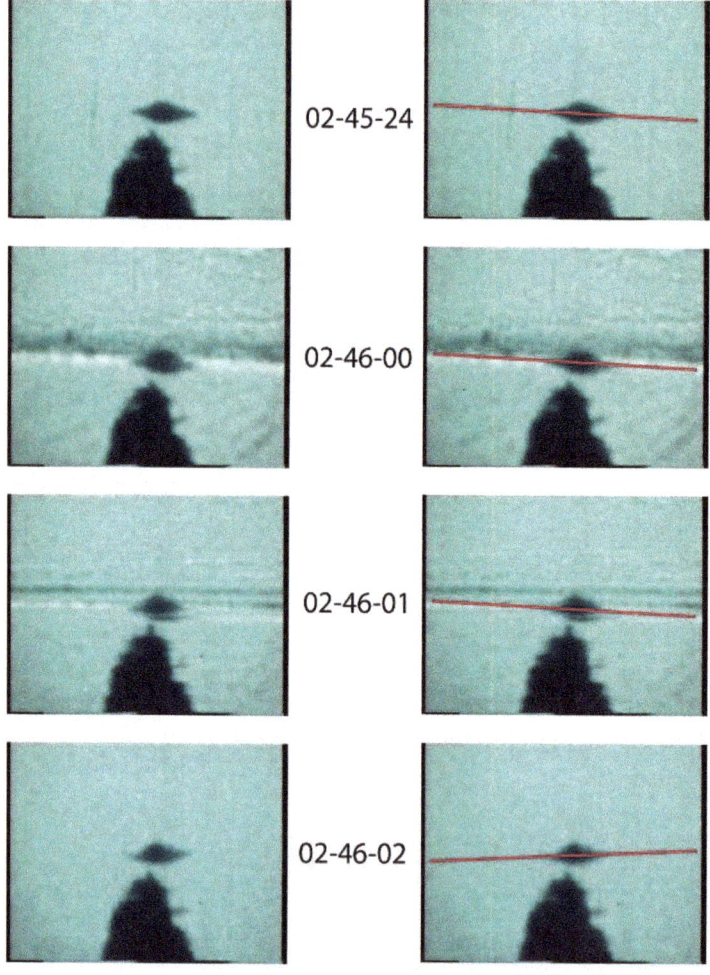

Figure 113 - Beamship flip. Left: frames from the PAL video. Right: An added line indicates the beamship's horizontal component, which rotates five degrees in less than 1/24th of a second.

The Beamship Demonstration

It is unlikely a coincidence that one of the four pulses happened at precisely the time of the flip. After the flip, the beamship moves backwards. No ordinary object of any size, whether a model or a big craft, moves in this manner. This object seems immune to inertial laws of movement. Figure 114 shows the expected activity compared with what we find in Meier's film.

Figure 114 - Left-hand sequence: Normal movement of a model hung from a cord. The model should start a gradual rotation while initiating a lateral movement. Right-hand sequence: The beamship in the film rotates (flips) in less than 1/24th of a second, and then moves to the back.

No known objects, from little models to big flying machines like helicopters, can flip so fast. If a model is hung from a cord and pulled or moved to the back, the rotation happens gradually in time with the gradual horizontal translation (Figure 114, left). What we see in the film seems to flaunt inertial laws as we know them. Presumably, the beamship has a way of overcoming them because it flips in less than 1/24th of a second and recedes from the camera horizontally towards the back.

The two middle frames in Figure 113 show the bright arc from the "pulse". We know it is just one single arc of light, but the PAL video interlace effect produces it in two consecutive frames. (Refer to Figure 89 to appreciate the effect). Let us suppose the bright arc from the "pulse" is produced in just one single frame marked "B" on the Super 8 original film (Figure 89, left). Two consecutive frames in the PAL video taken at 1/50th of a second show this bright arc ("B" detail) because two interlaced images capture the same frame in the Super 8 film when in PAL format.

The object's quick flip is yet another mystery to add to the film's litany of surprising hidden details.

The Beamship Demonstration

6- UFO Wobbling

Anyone keenly perceiving tiny differences and characteristics in a moving object soon notices an abnormal wobbling in the beamship, a wobbling that pre-empts a model hanging from a cord. Stevens describes *The Pendulum UFO* wobbling in his book *UFO Contact from the Pleiades – A supplementary investigation report* (page 240). In *The Pendulum UFO* film, the beamship wobbles in a very peculiar way. The clip from the *Contact* documentary shows how the beamship wobbles back and forth after the right turn. The unnatural wobbling is, however, not something unique to this particular UFO. It is easier to perceive it in the performance of other beamships in the same FIGU documentary. Anyone can watch one of three beamships in the sky wobbling in this particular way in the YouTube video *Demonstrationsflüge / Demonstration flights (IFOs, not UFOs)*. Also, watch here the beamship in a pink sunset sky emit flashes of light and wobble:

https://www.youtube.com/watch?v=EAHfOmvz6_s

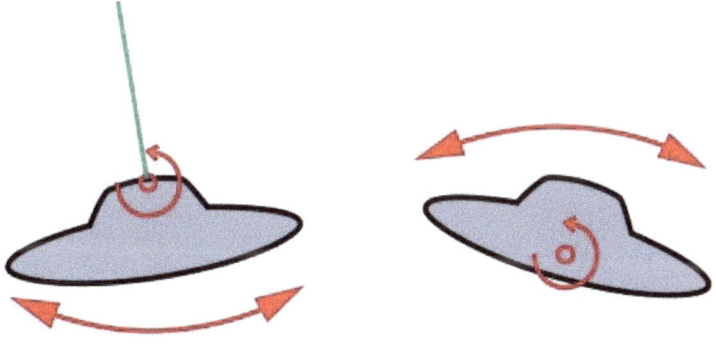

Figure 115 - A model hung from a cord wobbles around its top support contact point or its geometrical centre, but the beamship wobbling centres explicitly around a spot at the bottom of the beamship.

They are Here

Figure 115 illustrates the beamship's non-vertically hung pendulum wobbling. A model hung from a cord naturally wobbles around the support point where the cord affixes to the top of the model or its centre of mass typically at the geometrical centre. The beamship wobbling, however, centres explicitly around a point at the bottom of the beamship. This positioning could bring up the question of whether the supporting system has an anti-gravitational engine located inside the beamship close to the base.

7- The "pulses"

Perhaps the most extraordinary details, hidden in *The Pendulum UFO* film, are the "pulses". We call them pulses because they give the impression the beamship emits energy pulses that the film captures in a single frame. So far, we have detected four pulses, but perhaps there are more.

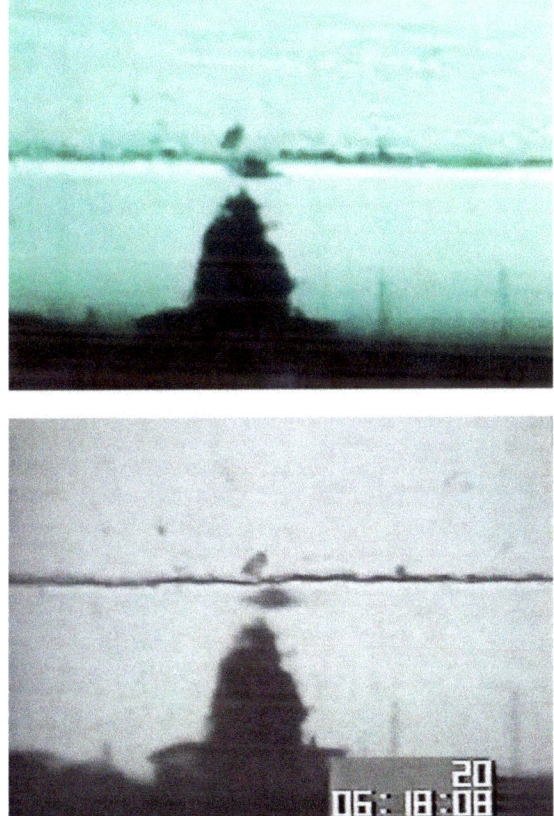

Figure 116 - First "pulse". Top: The PAL video. Bottom: The Nippon TV documentary.

It is possible to simulate a "jump" by cutting the film roll. Explaining the "pulses" as a trick, however, requires using a tool like Adobe-After Effects, that was unavailable in the 1970s. Trying to scratch or imprint the bright arc for the "pulses" on tiny film frames is exceptionally complicated, and it is not a single line, there may be several imprinted lines. Furthermore, the video was handed over to independent viewers, notably the Austrian TV team, and there has never been any mention of lines purposely scratched onto the film.

What are the "pulses"? One is in the Nippon TV documentary and the video used in our first investigation (Figure 116). We previously did not notice them, but after finding more and realising they correlate with the beamship's position, they are all too apparent. Figure 117 shows the four "pulses" we have found to date.

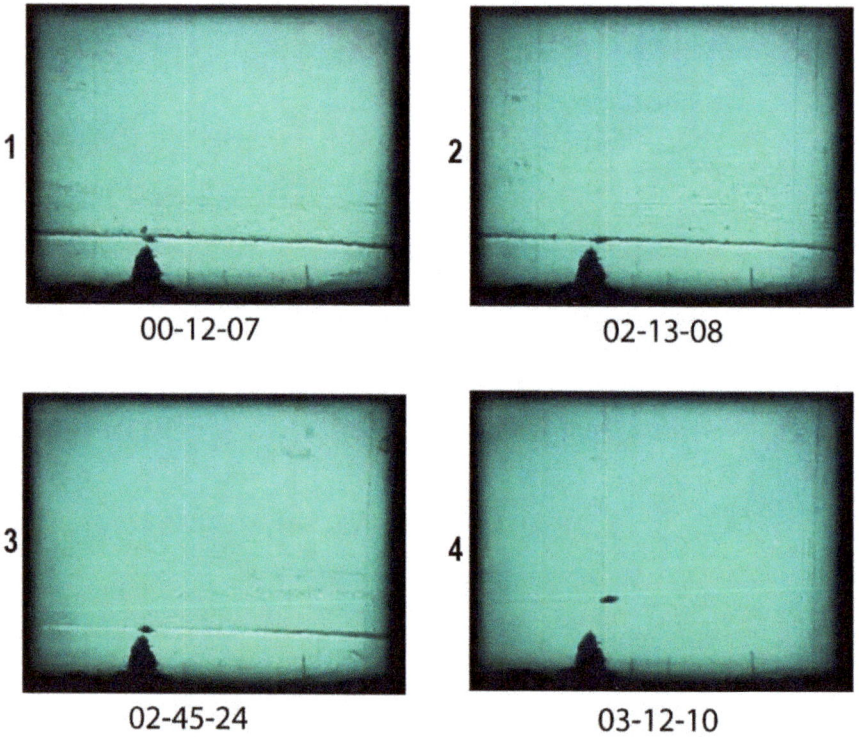

Figure 117 - The four "pulses" with their corresponding times in the film.

The first "pulse" happens when the beamship is motionless above the tree at the beginning of the film. Noticeably the bright arc is accompanied by a dark arc that crosses the beamship image, and there are distorted lines above it following the same direction. A piece of dust lies on the film very close to the ship that is also visible in the Nippon TV documentary.

The Beamship Demonstration

The second "pulse" happens as the beamship is moving back and forth between the second and third jumps. Again, a bright and a dark arc are visible together, and again, distortion arcs appear above them. These features are visible in just a single frame in the original movie. Whatever creates it is swift, 1/24th of a second or less. The arcs again cross the beamship image when it is in a similar position to "pulse" 1, but the UFO is moving.

The third "pulse" happens at the precise moment the beamship makes a quick flip, as explained previously. Again, the same arc of light that crosses the beamship image is present together with similar distortion lines above it.

The fourth arc of light is slight. We call it the "little pulse". Harder to find, there may be more of similar mild intensity in the film. It displays the same features, just less intense, and again, it crosses the beamship image.

Figure 118 - Fourth "pulse" or "little pulse" detail. The arched line of less intensity also crosses the beamship image.

If the film had only one of these "pulses", it could be a random film distortion. But this cannot be said for four of them, all showing the same arch, always crossing the beamship image, even when at a higher elevation. It is no coincidence that at precisely this frame, a "pulse" occurs and the beamship flips. Perhaps something is happening inside the beamship that produces this effect. Could a device emit these pulses, like flashes going away in the direction of the projected plane of the UFO base, and the pulses affect the film image in just 1/24th of a second? If so, in the fourth "pulse" the beamship is at a higher elevation, and so this projected plane must also be higher. Could this, in turn, cause this 4th "pulse" to be less intense?

Could the arcs on the film be faked scratches made purposely on the negative and the dark line the raised scratch portion of the film lying next to it? First, a faker is unlikely to deliberately scratch a line right through their UFO without ever drawing attention to the line; but a simple test belies this faked possibility.

Our modest purposely scratched line on a 35 mm camera film negative, Figure 119, shows no upper linear residue that creates any accompanying dark arc. While this 35 mm negative film is a little different from a movie film negative, it is highly unlikely so different as to account for an accompanying dark line above a scratch. Anyone can test a 35 mm film scratch for themself.

Our deeper and broader scratches raised the film emulsion right off the film surface and bundled it up below the end of the deep scratch. See, for example, the arrowed large grey bundle hanging down from the line on the lefthand side of Figure 119. Large scratches pull the emulsion off the film, and it also happens with some thin lines scratched on the film. The gelatinous film emulsion comes off in one slither; it does not rise onto an edge of the scratch to create a dark line but pulls off the film. With thinner scratches, the emulsion clumps into tiny balls lying at intervals along the white scratch line (see Figure 119, bottom scratch). Of course, these show no resemblance to the pulses in Meier's video.

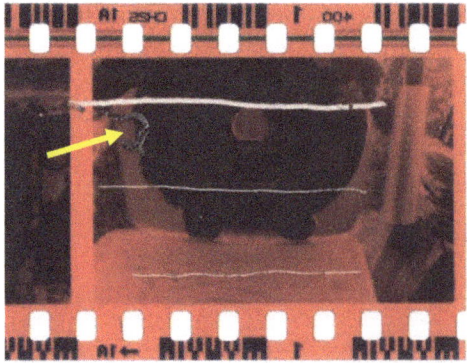

Figure 119 - The authors' purposely scratched negative showing no accompanying dark line for small or deep scratches, and both types pulled emulsion off the negative. Neither form of scratching produces an accompanying dark line.

Some of the dark "pulse" lines in the film are even clearly linearly separated from the light lines, like two parallel lines, one dark and one light. The raised emulsion curls up and off the film from scratching, revealing something quite different from what the film shows.

It is also surprisingly tough to make a very straight scratched line as in the film even when using a ruler or guide.

Finally, even if this dark line were the raised emulsion going along the whole length of a scratch its tone here is way too dark. The raised emulsion piece is from the "light" sky and not being doubled over it would be expected to be only twice as dark as the light sky whereas it is generally orders of magnitude darker along the whole length. Something decidedly different from scratching the film negative is going on here.

They are Here

8- Electrostatic flashes

We have found a white band on the top of the frames when the beamship jumps. We also found arcs of light during the "pulses", and the camera also jumped during the UFO "jumps". There must be some induced electricity inside the camera, perhaps created by the beamship. We show more evidence of electrostatic charges here.

That electrostatic charges inside a camera can imprint a mark on photos and videos is well known. To complete the list of interesting and sometimes unexplained features in *The Pendulum UFO* here are some electrostatic charges that we have found.

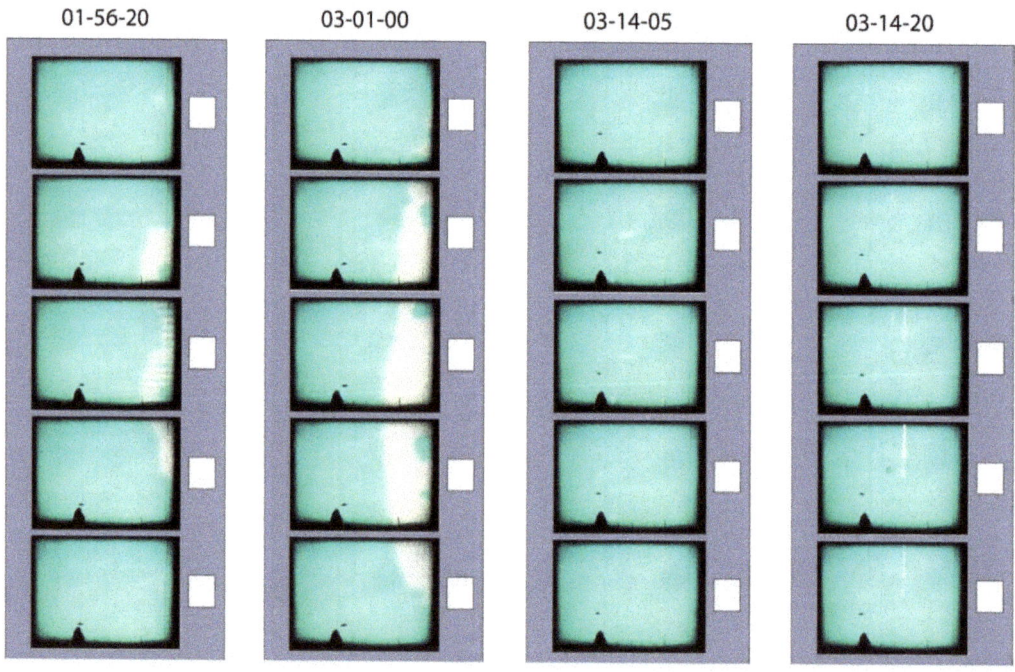

Figure 120 – A composition of four examples of internal light charges inside the camera with the times they happen in the film.

Figure 120 is a composition of four frames that present this "lighting" event in the film. We excluded some interlaced frames in the second one to make it no more than five PAL video frames.

We see the continuity in some of them, like the second event, that lasted for several frames.

What happened on this snowy afternoon? The evidence indicates a well-prepared demonstration by a mysterious flying device performed in a way that somewhat mimicked a pendulum but left hidden clues to discover, after years of protracted debate, that this is not just a simple model hung from a cord. The evidence in this Billy Meier beamship demonstration video contributes to the worldwide UFO controversy. It is also, however, a handy tool for ushering in a smooth change from ancient anachronistic beliefs and "knowledge" to a realisation that *They are Here*.

Conclusions

- Even without the original film, the several copies available give us sufficient information on what happened in this place close to Hinwil on a March afternoon in 1975. The original film right now could be severely damaged, but the copies preserve the nature of this demonstration. It would be superb to find and study at least one good copy of the film in Super 8 format to continue investigating it and perhaps find more hidden details. Every time a new copy does surface, more fascinating details emerge.

- It is essential to know the recording system used, be it PAL or NTSC format, and how it works.

- Since time measurements in Meier's video film show any hypothetical pendulum length must change at many points in the film, we conclude that a pendulum model version would require a sophisticated 11-metre-high arrangement to explain most events occurring in *The Pendulum UFO*. Yet, even with this "sophisticated" difficult-to-achieve model, we still cannot explain several exciting features found in the film.

- No model on a pendulum theory can explain all of the films' events and features. We note just eight critical features of this film:

    1. The UFO moves in a forced pendulum movement but does not follow the simple laws of pendulum physics. Several attempts to replicate the actions have failed because while they show something somewhat similar to the original movie, they lack precise representations and explanations of all these hidden details. The actual variations in the pendulum movement have never been replicated.

    2. The images show the tree to be a large one, close to the house. We know it is not a little tree. Evidence shows the tree existed, although now it does not. A

little tree close to the camera would look completely different from what we see in the video

3. The UFO performs a unique smooth sharp turn, which cannot be simulated with a model because the model always shakes. The beamship moves the treetop without touching it, something a little model cannot do.

4. The UFO performs three "jumps" in space which burned a bright band at the top of the film and somehow made the camera jump, which is far more indicative of unknown energy interaction.

5. The beamship makes a 1/24th of a second flip, which is impossible for a model or any conventional craft.

6. This beamship wobbles as if supported from a bottom point, not like a model supported from a top cord.

7. The beamship appears to emit four "pulses" captured in single frames (four times), and the pulse appears as a continuous, wide bright arc accompanied by a continuous dark arc that always crosses the UFO image, even when the UFO is at a higher elevation.

8. There is ample evidence the camera was subject to electrostatic charges, which opens up the question of whether the beamship produced them.

- As with other pieces of evidence in the Billy Meier case, the familiar pattern emerges at first glance. It seems natural to conclude that the demonstration could be a hoax. However, upon detailed investigation, hidden clues arise indicating the opposite and that the film is most likely what Meier says it is: a film of a sizeable extra-terrestrial UFO.

- The manner of the demonstration invites debate, thus achieving a general worldwide UFO controversy. To us, however, there is no longer a controversy: *The Pendulum UFO* film shows a flying ship of around seven metres performing technologically in a manner unknown to us, including jumping in space.

They are Here

- *Independence Day* special-effects award winners Engel and Weigert agree that "it would have been very hard, probably even impossible, to fake this kind of shot."

Practical Pendulum UFO Experiments

Introduction

The Pendulum UFO shows a dynamic object. No matter it could look like a toy hung from a cord, it is a significant object imitating pendulum oscillations. In the previous chapter, we demonstrated how difficult it is for a crew of people to reproduce this film of Meier's. We also discussed several features showing this object is flying unlike any machine usual to Earth, like the three "jumps", the flip, the "pulses", and others.

The Pendulum UFO video, as indicated before, is readily available on the Internet. Even old documentaries contain the film. We are sure the film has not been altered to change the object's speed because we see the same pendulum period disparity in every copy of the film, taken at different times, so the variation in the oscillation periods are real and indicate the length of a hypothetical pendulum that must be changing all the time, which as we discussed, is extraordinarily complex to emulate. So much so that no one has ever done it.

In this chapter, we suggest a practical way to experiment with pendulums. First, however, as an introduction to any such test, we encourage the watching of Zahi's (Francisco's) YouTube video *OVNI Danzante - Dancing UFO* that includes the investigation details. The link is:

www.youtube.com/watch?v=IKeutVKFbG0

Finding a computer tool

To accurately measure a pendulum period, it is recommended to record it on film and check it frame by frame. Then to experiment with pendulums, simulating the Meier film, and verify the period variations in the original movie.

Several video editing tools are available. The tool selected should allow images of every movie frame. Adobe Premier, Adobe

They are Here

After Effects, Pinnacle Studio, Final Cut, Media Composer, among others suffice. The video can even be loaded and viewed in Windows Movie Maker. Figure 121 shows the application screen.

What follows is an explanation of how to use the free Windows Movie Maker available for PCs. Similar tools exist for MACs.

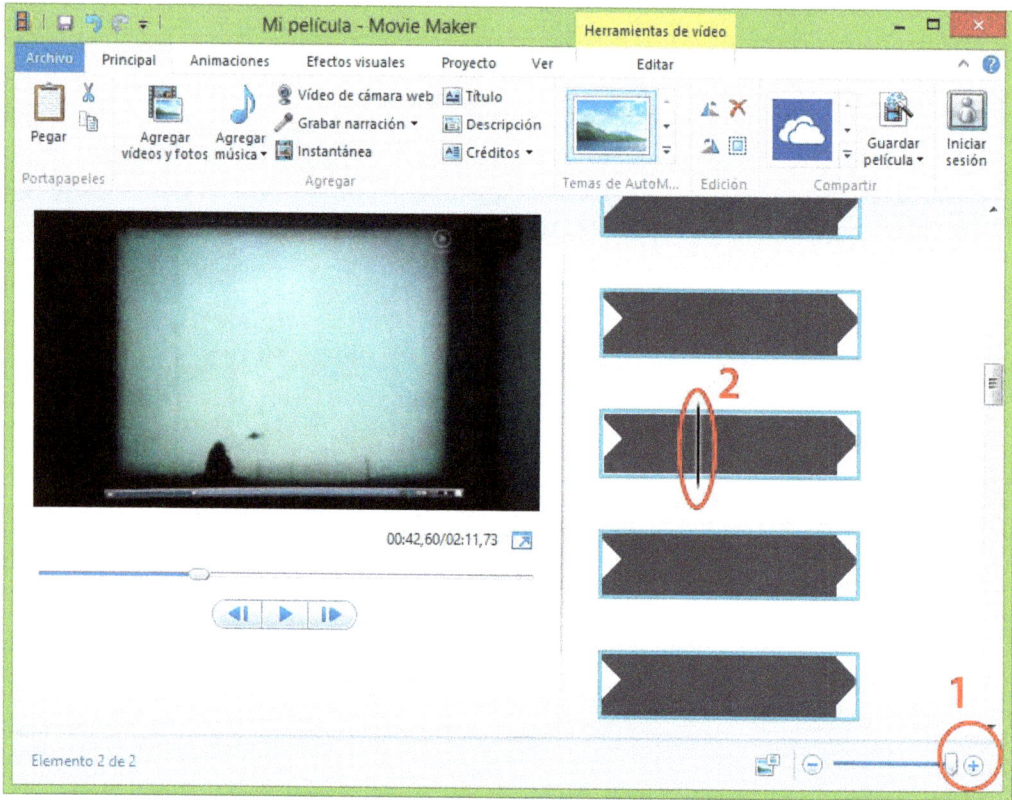

Figure 121 - Windows Movie Maker screen.

To exemplify the process, increase the timescale zoom level to maximum (**1** in Figure 121). Then move the black line bar **2**, which advances the movie slowly frame by frame. Below the current video image, the time measurement indicates the time of the current frame viewed, and the total video clip length.

Next, move the video cursor until the UFO reaches a specific position (a reference point), like the right or left position, or a spot just above the tree. Then advance the video until the UFO returns to the selected reference starting point. Take note of the time

Practical Pendulum UFO Experiments

measured below the video image at the start and endpoints. Finally, subtract both values to obtain the *period*.

Now you can record your experiments with a camcorder or smartphone and review them with your Video Editing tool selected.

Experimenting with pendulums

Find a surrogate UFO model. A pot cover or any small solid object works fine. Attach it to a cord and play with different lengths. Note down the pendulum length – the distance between your surrogate UFO's centre of mass and the node (where the cord is attached).

Hang your model on a high branch or use a long pole like a fishing rod. Try different lengths recording yourself moving your model. Then use the video editing tool as described before and measure the different *periods*. Does the *period* change if you have the same pendulum length? Does it change if you move the fishing rod imitating planar, conical, or spiral pendulum movement? (Refer to Figure 91 to see each type of pendulum movement.)

Once you measure the *period*, use it to calculate the cord length, and compare it with the length used in your experiment. Use the following equations to find the pendulum length based on the period:

$$L = 0.2482\ T^2 \quad \text{(in metres)}$$

Or,

$$L = 0.8144\ T^2 \quad \text{(in feet)}$$

Where **T** is the pendulum *period* (in seconds), and the resulting length, **L**, is in metres or feet depending on which formula you use.

This equation comes from the pendulum formula found in any physics book:

They are Here

$$T = 2\pi\sqrt{\frac{L}{g}}$$

Where **g** is the acceleration of gravity.

Using this pendulum physics formula, we can know the pendulum length by measuring the pendulum *period* (time to complete one cycle).

The Pendulum UFO movie Meier recorded is available for download at any of the links previously provided. Measure the UFO movement *periods* to find the length of the "cord" with the formula above.

Challenging with a pendulum UFO

If you are curious to know whether Meier filmed a small UFO model or a large object you could try the following proposed "challenges" that inform how easy or difficult it is to replicate Meier's film.

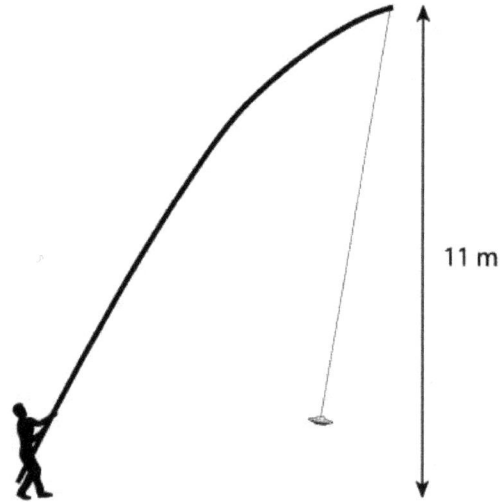

Figure 122 - Testing with a pendulum.

Practical Pendulum UFO Experiments

Challenge 1: Using a long pendulum with one hand.

If Meier used a pole with a one-foot diameter model UFO attached to it, anyone could do the same. As indicated before, you could use a long pole, like a fishing rod. Refer to the pendulum lengths in Table 5. In phase 12, the pole must rise very high to maintain the same period. We calculated that a 15-metre-long pole is needed (49 feet) because to raise the model as in phase 12 while retaining a pendulum length of over 5 m (16.5 feet) the pendulum node must be 11 m (36 feet) high. So it is best to find a pole this long to try reproducing the measurements (Figure 122). If the length proves impossible, try a 10-metre-long pole (33 feet) and climb up onto a platform five metres (16.5 feet) above the ground.

Perform the planar pendulum movements, then the conical (circular) motions and finally raise the pole until the model is six metres (20 feet) above ground level. Try with both arms, and then with only one arm, since Meier has only one arm.

Challenge 2: Changing the pendulum length

Our second simulation challenge relates to phase 5. The UFO moves in circles around the tree in this phase and, in three consecutive circles, the *period* changes from 4.4 to 5.1 seconds, and then lowers to 4.3 seconds. So the pendulum length changes from 4.8 m to 6.5 m and then back to 4.6 m. Therefore, the pendulum length changes 1.9 m (6.2 feet) in five seconds.

The challenge is to replicate this. With a fishing rod or tree branch above, move the model in circles, and in two consecutive circles, change the length from 4.8 m to 6.5 m and then to 4.6 m in the third circle. The UFO model must not move up or down but always remain at the same elevation. Record it and compare your result with Meier's film.

Using a fishing rod support enables speedy pulling and extending of the cord. With the model supported by a tree branch, it is a formidable challenge to change the pendulum length by 1.9 m because the tree branch must lower by 1.9 m.

In Langdon's test, he hung his model from a branch in a nearby tree and, as indicated in his video, moved the model in circles obtaining the same period of 3.72 seconds in each of two circles, which is equivalent to a pendulum length of 3.43 m.

Meier's air pendulum length, as we pointed out, was variable. Langdon's test needs challenging with a higher tree to give a pendulum length of around 5 m.

Challenge 3: Moving the treetop without touching it

Right before the UFO makes a sharp turn, it is evident the treetop moves. This challenge is an attempt to simulate the treetop movement or bouncing. Using a little tree like a little cypress pine of around 1 to 1.5 m in height, move your UFO model above it without touching it. The little treetop must bounce as in the original film. Record the experiment on a camcorder and compare it with Meier's film. If the model somehow moves the little treetop, ensure the model does not touch the tree or shake and that it maintains a smooth movement as in Meier's film.

Challenge 4: Performing a right turn without shaking

After the treetop movement, the UFO makes a 90-degree turn. Simulate this phase. Attach an extra cord if needed and pull it. The model must not shake or wobble. Compare your film with Meier's and notice how smooth his turn is. Try to replicate it.

Challenge 5: Creating the UFO blurriness and a little tree

Using a model UFO and a little tree, simulate the UFO's difference in blurriness when farther away compared to when it is close to the camera.

Place the little tree at around 17 metres from the camera with a distant house in the background. Ensure the environment is a bit hazy. Record your test and confirm whether it looks like the original film or not. Do the tree and house have the same degree of blurriness and the UFO more blurriness when farthest away?

Challenge 6: Creating the wobbling

Do a test with a model, supporting it from a point close to its bottom. Allow it to shake and confirm the rotation point. Record your findings.

Challenge 7: Simulating the pulses

Find a way to simulate the bright and dark arcs seen during the "pulses". Ensure they last less than 1/24th of a second. Try using a flash. Compare the results with the original film.

Challenge 8: Shooting the whole scene in a single take

The last challenge is to record the 12 phases in a single shoot. The camera must remain on during the entire performance. Do not record several clips and later join them together. It is one single take lasting more than three minutes. Use as many assistants as required. Any arrangement is suitable: the "fishing rod" support, the "above tree" support, or our suggested support of using an elevated platform.

By performing these challenges, you can know if a single human being, with just one arm, could create this film with a scale model and little tree. Alternatively, you may realise the filmed object is not a little model but a significant object, or beamship of seven metres imitating a pendulum.

Part III

The Hasenbol Beamship Demonstration

Encounter with a New Beamship

Introduction

Part III discusses the details of an extraordinary demonstration supported by abundant evidence from the Plejaren. It occurred on 29 March 1975 at around 5 pm and lasted for more than two hours in Hasenböl-Langenberg, close to the town of Fischenthal in Switzerland. We refer to it as The Hasenbol Beamship Demonstration. According to Billy Meier, Semjase accompanied by Quetzal brought a new type of beamship capable of inter-dimensional travel. Stevens indicates they landed and talked very briefly with Meier (*UFO Contact ... – A preliminary...*, page 343).

Figure 123 - Meier's photo #174 of 29 March 1976 at 6:02 pm. The new beamship is in front of the sun to the west and behind a tree.

When observing Meier's photos, the difficulties he endured to reach the designated contact site usually escape us. To witness the Hasenbol beamship, he climbed 260 metres above the surrounding terrain to the top of a hill, which Stevens describes as difficult to reach at the time even in a four-wheel-drive vehicle. Yet Meier sometimes had to push his moped while pulling a little trailer carrying his cameras and equipment all with one hand *(UFO Contact ... A preliminary..., page 345)*.

Meier's photo #174, Figure 123, relates well with a scene from the Nippon TV documentary video discussed later. The tree close to and in front of the beamship is no longer there, only its stump remains, perhaps after being cut down to open space for what appears to be a planned future crop area on the hilltop.

Four pieces of evidence, namely, Meier's photos, the Nippon TV documentary, a video produced by FIGU in 1998, and recent satellite photos (2018) available in Google Earth, confirm vegetation growth and the location of other trees on this hill. The excellent FIGU documentary on the Hasenbol beamship is freely available here:

https://youtu.be/vq8PaVsJB6Y

On this early evening, Meier shot at least two rolls of film with his Super 8 mm movie camera and took more than 100 photos. For unknown reasons, two rolls of 36 photos never returned from the photo lab and so are lost. No one knows what was on these rolls, although people close to Meier think they probably contained the last part of the encounter when the Plejaren went back to the sky.

Remarkably and unfortunately, most of the many photos Meier took have been stolen, lost, or strangely disappeared, not only the Hasenbol ones. Also, more than a few alterations were attempts to discredit Meier, and it is not surprising to find lost photos in this event.

What we, the authors, do in this Part III can be done with any alleged encounter between Meier and the Plejaren space beings. It is not necessary to visit Switzerland to investigate this case. It is both fascinating and revealing enough to explore by studying and researching available good quality photos, videos and information in the several books available of Meier's photographs.

With the help of excellent tools like Google Earth, and aerial photos (Swiss Confederation) we can situate ourselves exactly where Meier had his encounters, calculate details and distances, and then using these calculate with surprising accuracy the beamships' movements during their demonstrations. An analysis similar to this Hasenbol beamship analysis is possible for other Meier UFO encounters. For example, *Fuchsbüel-Hofhalden* with a background lake where the beamship flies around a tree that the Plejaren later caused to "disappear." Or the occurrence at *Bachtelhörnli / Unterbachtel* with three beamships flying above a distant valley, or any other demonstration.

Hasenbol location

Figure 124 - Clip from Google Earth. In the upper section, BM marks Billy Meier's encounter location on a hilltop east of Fischenthal town.

Billy Meier's location during the demonstration is marked BM on the Google Earth August 2018 satellite photo, Figure 124. The town of Fischenthal is west of Meier's position. The hill is 1,000 metres above sea level and the town at 740 metres making the hill

260 metres higher than Fischenthal. The Google Earth location is Lat-Long coordinates 47° 19' 49" North and 8° 56' 06" East.

Several details observed in Meier's photos and Super 8 mm camera videos, the Nippon TV documentary, and the FIGU documentary, are visible in old 1972 aerial black and white photos and recent satellite photos from Google Earth (Figure 125, top and bottom). Some trees have grown while others have been cut down or logged. The *large-tree* in front of the beamship in photo #174, and other trees behind it are missing today. To better acquaint ourselves with the demonstration and filmed area, we now briefly describe Meier's location and the photographed trees which we have numbered **1** through **6** in Figure 125.

Number **1** indicates Meier's location in most of the photos and videos during the contact demonstration.

Number **2** is the now missing *large tree* standing prominently in front of the beamship in Meier's photos, west of his observation point (the top is North in Figure 125). This *large tree* plus other trees were probably logged for crop space by a farmer, based on satellite and recent aerial photo indications of crop activity. However, the Nippon documentary, Stevens' books, and 1972 Swiss Confederation aerial photos (Figure 125, top) confirm the tree's size. A *little tree* (Figure 127) previously proximal to the *large tree* has also been logged. Notice the differences between the trees in the 1972 aerial photo in Figure 125 top and the recent Google Earth satellite photo in Figure 125 bottom.

Number **3** is the *south tree* downhill. Being downhill only its upper part is visible behind the *large tree* in photo #174. It appears in Meier's photos, the Nippon TV documentary, and the FIGU Hasenbol documentary in 1998, but it too no longer exists.

Encounter with a New Beamship

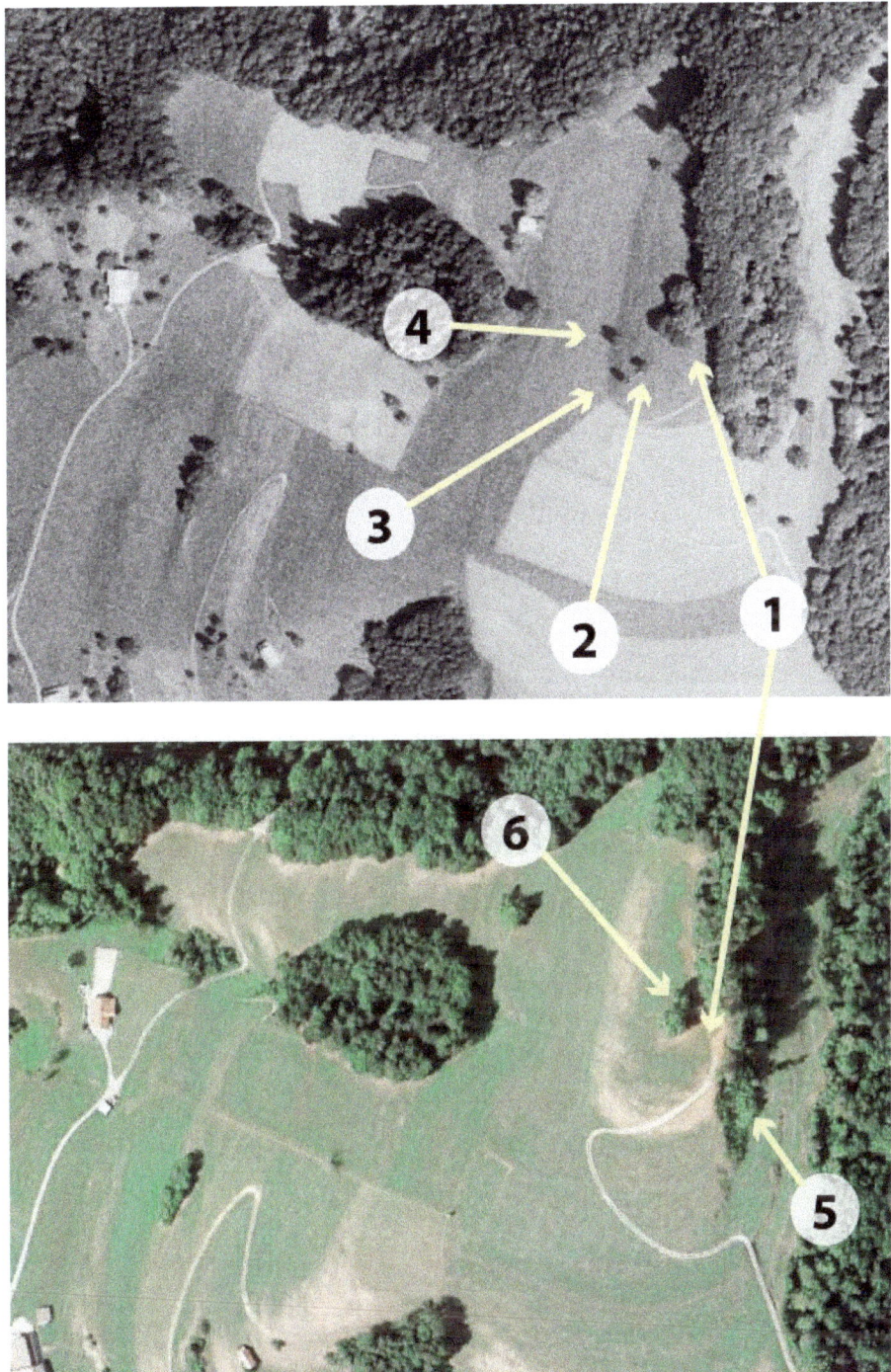

Figure 125 - A 1972 aerial photo of the Hasenbol location (top) compared with a 2008 Google Earth image (bottom). **1** marks Meier's location. Trees **2**, **3** and **4** in Meier's photos, along with some other trees, have subsequently disappeared.

They are Here

Number **4** is the *north tree* that was also downhill and partially visible to the right of the *large tree*. Only its upper part peeped above the hill on the right of photo #174 (Figure 123). It too has been logged so no longer exists and is absent in recent satellite photos.

Number **5** is the *eastern woods*. The left side of some photos probably show some tree branches from here, or perhaps all of the branches seen on the left in photos #152, #153, #154, #155, and #181 are from the *north grove*.

Number **6** is the *north grove*. Meier installed his movie film camera at some point in time behind or within these trees and recorded the beamship moving east to west and west to east with his camera pointing towards the south and Aurüti mountain, the south snowy mountain seen in both the Nippon and the FIGU Hasenbol documentary video scenes. In one FIGU video clip, branches sway in the wind in front of the camera with the south snowy mountain in the background. A second video clip from the same documentaries shows the beamship flashing pulses of light. This time, the camera points towards the south-west in the direction of Fischenthal. To film this Meier most probably positioned himself very close to the mark **BM** in Figure 124.

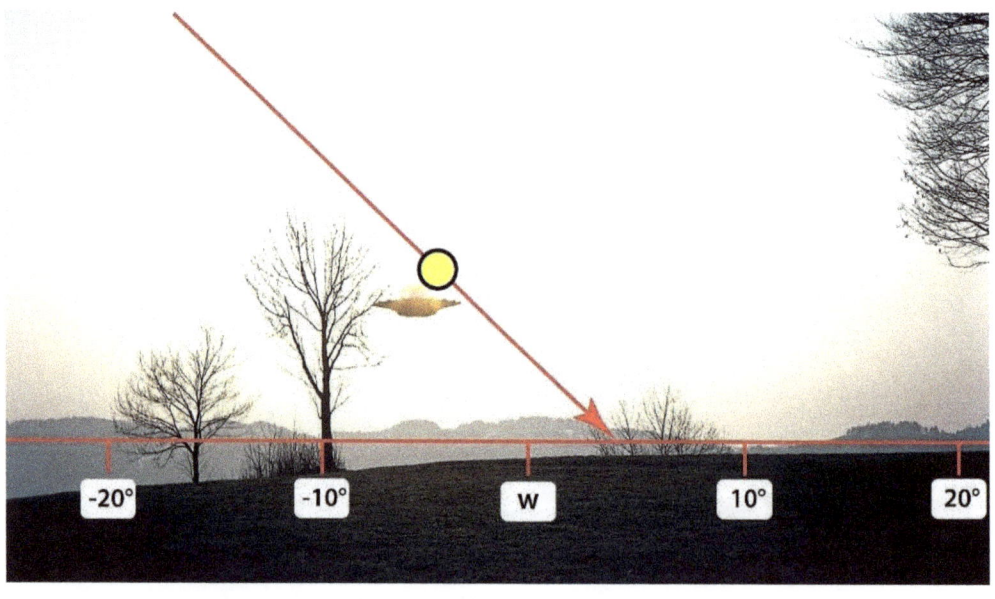

Figure 126 - Estimate of due west **W** on photo #174.

Encounter with a New Beamship

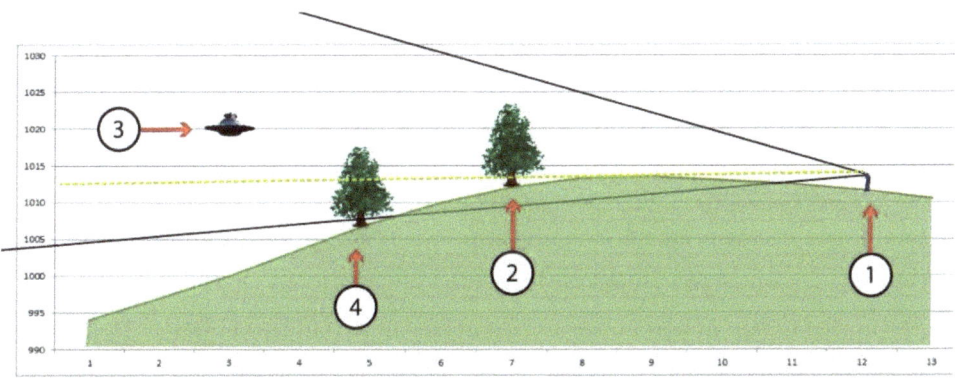

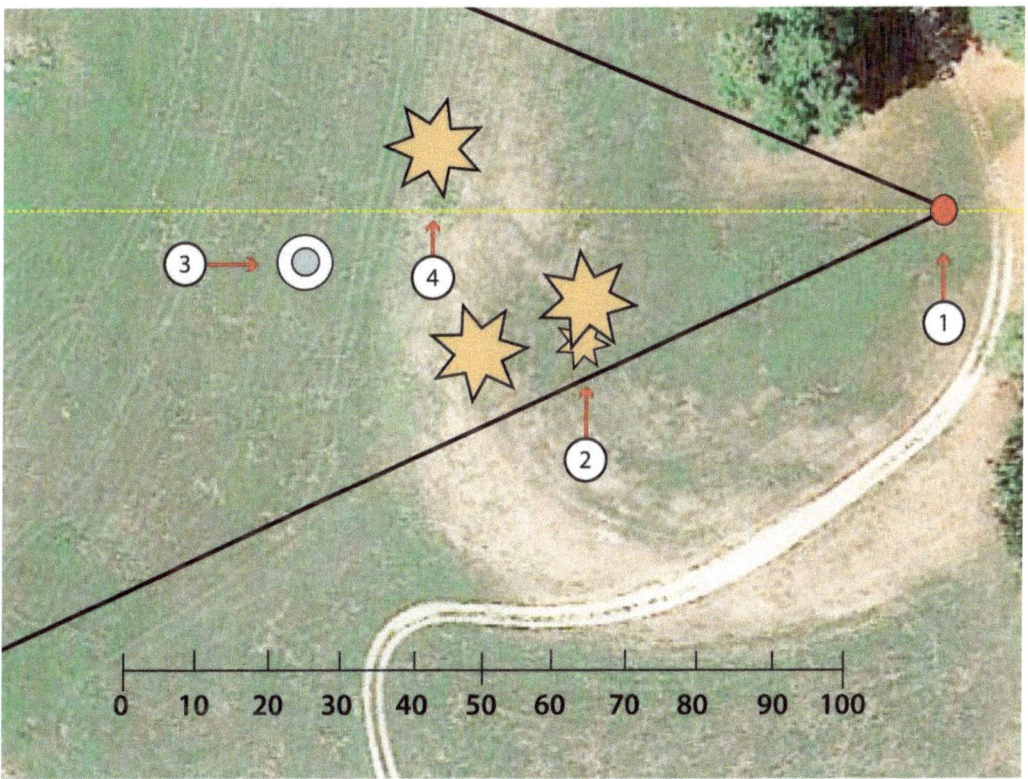

Figure 127 – Horizontal terrain elevation (top) and bird's-eye plan view (bottom) for photo #174. Elevation and horizontal scales are in metres. Terrain contours and photo from Google Earth. Number **1** is Meier's location, **2** is the *large tree* with the *small tree* under it, **3** is the beamship, and **4** is the *north tree*. The *south tree* is the large unnumbered orange star. North is to the top.

In the Nippon TV documentary, Meier takes Yaoi to the hill and demonstrates how he set his film camera on the tripod that he

They are Here

used. The documentary shows Meier filmed from close to the *north grove*, and most of the time he was probably around the **BM** area.

Meier's Olympus camera has a 46° horizontal angle of view, as shown in Figure 126, together with 10-degree increments toward the north and south from due west **W**. The *large tree* in Figure 126 (and **2** in Figure 127) is 10 degrees south of due west. Its base is two degrees below the horizon, almost at Meier´s feet level. The beamship is seven degrees above the horizon.

The 1972 aerial photo (Swiss Confederation) in Figure 125 (top), helps us locate the four trees missing in current satellite photos: the *large tree* (**2** in Figures 125 and 127), the unnumbered *small tree* close to it probably too small to see (marked with the little star in Figure 127), the *south tree* (**3** in Figure 125 and marked with the large star symbol west of the *large tree* in Figure 127) and the *north tree* (**4** in Figures 125 and 127).

Figure 127, top, offers a horizontal terrain elevation for the dashed yellow line (East-West line crossing Meier's location) on the bird's-eye plan view (Figure 127, bottom). Elevation and horizontal scales (top) are in metres. Terrain contours are from Google Earth, and the top and bottom figure numbers and positions correspond. Figure 127 numbers denote: **1** Meier's estimated location at 50 meters east of the *large tree,* which is number **2**; **3** the beamship position in photo #174, and **4** the *north tree.*

We find Meier to be 50 metres from the *large tree* by using the camera formula, estimating the tree's size from the Nippon documentary, and measuring its size in photo #174. The beamship is 90 metres from Meier in this photo. The following section explains this distance calculation, as well as distances in other photos.

Now located at the site, it is easier to reconstruct how Meier took all these photos at this isolated place and to show the beamship's movements during this eventful early evening.

The photographic sequence

A sequence of over a dozen photos of the Hasenbol beamship exists. The book *Photo-Inventarium* beautifully illustrates most of them (Meier, pages 80 ~ 90) and the *Future of Mankind* website gallery also exhibits excellent high-resolution copies.

The sequential photos show details of the large tree, little tree, north tree, south tree, the north grove, the eastern woods, and the distant south snowy mountain, among other features. The features indicate the camera's orientation and in some cases, Meier's location at the top of the hill, which changed as he moved around the mark **1** in Figures 125 and 127

Verifying in which direction a photo was taken, by looking at the distant mountains and nearby trees, helps locate the beamship's relative orientation at any moment. Still, we also need to know the distance from the camera. For this, we can use the camera formula previously used in the WCUFO analysis in Part I (*Below the WCUFO, with Trailer and Generator*). The formula is a geometrical association of four aspects: the focal length of the camera lens; the distance from the object (the beamship) to the camera; the size of the object on the film's surface; and the real size of the object observed (beamship diameter). Knowing three of these enables calculation of the fourth.

We know Meier used his Olympus camera with a 42 mm focal length lens, and we can also measure the beamship size in each photo and calculate how large (in millimetres) this image is on the negative film. Furthermore, based on Meier's claim that the beamship measures seven metres, its distance from the camera can be estimated and plotted on a map. Calculations for the idea that Meier used a little model of 55 cm or 40 cm are also possible, and the next section covers that discussion.

Camera formula calculations for the distance to the beamship in each photo gave the results in Table 9. This calculation required assurance that the photos were full-frame images, not cropped, image. Some cropped photos appear in the *Photo-Inventarium* book, but full-frame pictures are available on the *Future of Mankind* (FoM) website. So when necessary, we used the FoM versions. For some pictures, we had to look at several photos and make a best-estimate of the full-frame size. Asterisks denote estimated photo distances for non-full-frame views, or when uncertainty existed as to whether the photo was full-frame or not. So in those pictures, photo distances could be longer.

Pic #	Time	Dist. (m)
151	17:16	293
152	17:18	264
153	17:20	252
154	17:21	388
155	17:23	384
157	17:26	76
168	17:50	60*
171	17:56	75*
174	18:02	90
175	18:04	88*
164	18:05	87*
176	18:09	59
179	18:09	55
181	18:20	50

Table 9 – Camera formula data for Meier's photos studied, with shoot time and camera distance in metres. * marks an estimated value.

Figure 128 plots all of these analysed photos. From a total of about 100 taken, Meier only received some back from the photo lab, and in this analysis, we show only 14 of those that came back. Little white circles in the bird's eye view of Figure 128 illustrate the beamship's size drawn to scale. North is to the top. The red dot close to BM marks Billy Meier's location, and the star symbols towards the west mark the four trees now missing. Immediately noticeable is how close the beamship was to the camera in nine photos, which explains the extraordinary details captured.

Three photos exist of the beamship in the sky behind the *large tree*, #164, #174, and #175 (Meier *Photo-Inventarium*, page 87). Comparing available photo #164 copies with the apparently full-frame or almost full-frame photo #174 copies reveals the #164 copies cropped. Noting the respective sizes of the beamship, *large tree,* and the mountains in the background enable distance calculations in #164. We found #164 shows the beamship image size is 3.8% larger than in #174.

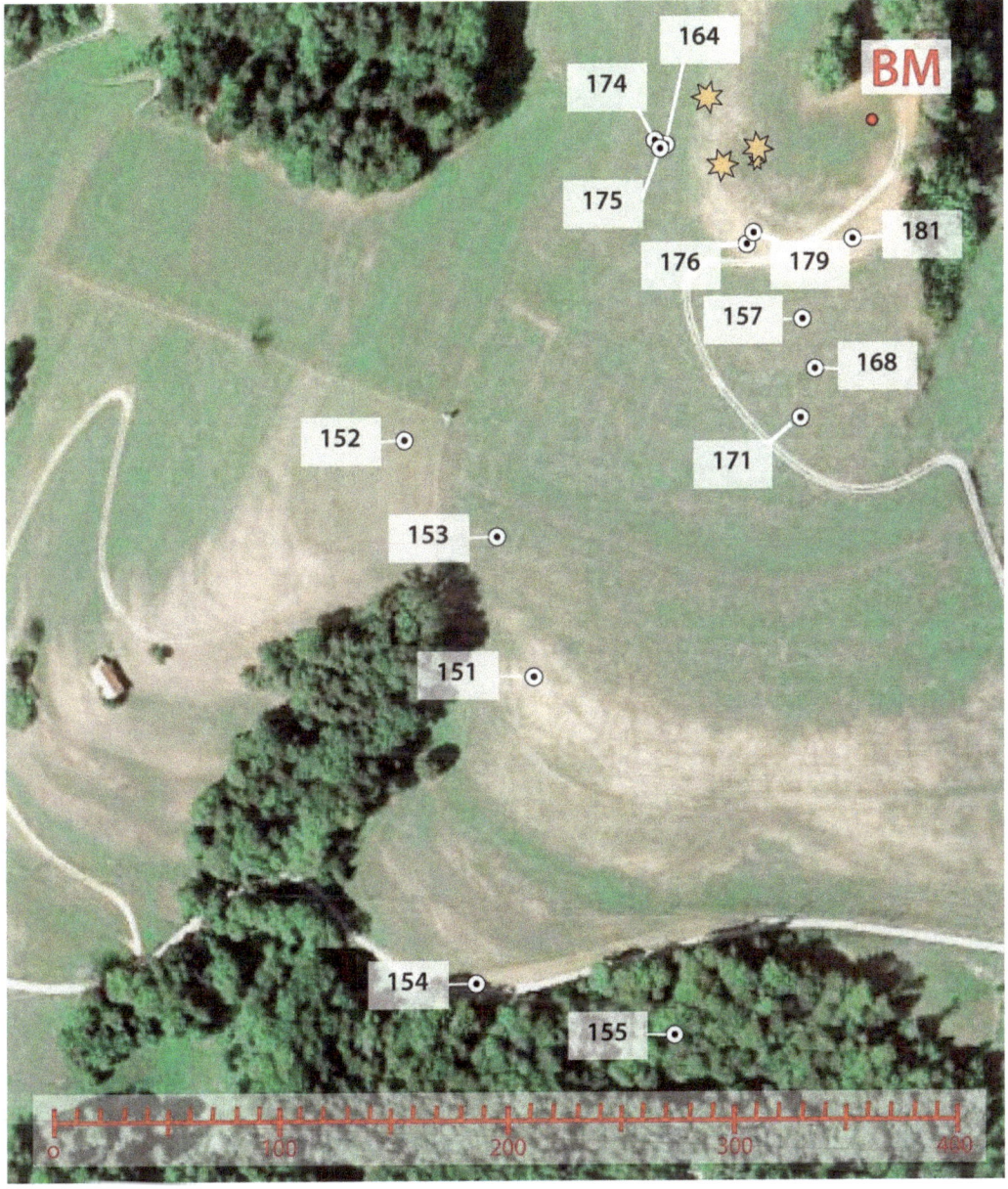

Figure 128 - Bird's-eye plan view, with the approximate beamship location labelled for each photo. North is the top.

They are Here

In the Nippon TV documentary, Meier says the beamship was just 5 to 10 metres behind the *large tree*. We determine it was farther. We calculate the *large tree* at 50 metres from the camera and the beamship 87 m from the camera making the distance between the beamship and the *large tree* in #164 37 m (87 m minus 50 m to the tree). Perhaps some of the missing or stolen photos from this series, however, show the beamship at just 5 m behind the *large tree* as Meier reported.

Maybe some of these missing or pilfered photos show the beamship considerably closer to the *large tree* and in even greater detail, perhaps clearly behind some of the *large tree* branches providing proof of its large size, making them more likely candidates for confiscation by interested parties. A secret agency wanting any of these photos would surely take the most revealing and instructive ones. Alternatively, perhaps even the Plejaren arranged for this *proof* to disappear.

These compelling photos that Meier took in the 1970s show a real flying machine. Yet, again the same pattern in Meier's evidence arises. Even this demonstration allowed sceptics and doubters to maintain personal beliefs that the case was false. The available evidence at the time required specific knowledge of things like Mie light scatter and precise calculations unknown to the general public. So for many people, it was inconclusive. Now, however, after image processing with tools available to anyone like Gimp or Photoshop, the evidence is conclusive (Zahi and Lock *Researching a Real UFO*). Today, with these tools we can readily confirm this beamship is behind the tree, at least in photos #164 and #175; and other evidence presented earlier in this book shows Meier's UFOs and beamships cannot be tiny models hung from a cord.

The following photographic sequence of the 14 photos plotted in Figure 127 shows and discusses each one in more detail.

Photo #151: 5:16 pm, estimated beamship distance 293 metres.

Figure 129 - Photo #151 towards the SSW. (Source FoM.)

Figure 129 is a full-frame high-resolution photo from the Future of Mankind "Photo Gallery". Meier most probably stood close to the *north grove*, some branches of which protrude in the top left corner or they are branches from the *eastern woods*. The *little tree* is at the far right. A small portion of the south snowy mountain is in view at the bottom far left as Meier aimed his camera towards the SSW.

The dark frame with rounded corners at the edges of this photo hints that it is full-frame. To calculate the beamship size on the original negative film requires comparing it with the full-frame size (36 mm wide). At seven metres in diameter, the beamship is 293 metres from the camera (camera formula data, Table 9).

They are Here

Photo #152: 5:18 pm, estimated beamship distance 264 metres.

Figure 130 - Photo #152 towards the SW. (Source FoM.)

Two minutes later, Meier took another photo (#152). This image is also high-resolution from the *Future of Mankind* (FoM) website. Now closer, the beamship has moved SW in the direction of Fischenthal, and the *little tree* and *large tree* at the far right are in better view. From here, it seems Meier is standing inside the *north grove* that today has fewer trees. Previous photos might also have been taken from within this grove. At this time of year, end of winter and early spring (March 29), most trees here still have no leaves.

Encounter with a New Beamship

Photo #153: 5:20 pm, estimated beamship distance 252 metres.

Figure 131 - Photo #153. The beamship has now moved back to the left. (Source FoM.)

This high-resolution photo #153 from the FoM website taken a couple of minutes later shows the beamship now having moved to the left, to a location between #151 and #152 as it comes closer.

Meier remains standing behind or just in the *north grove*. A different type of tree (a fir?) in the bottom half of the picture is close enough to the camera stuck on infinity to be out of focus. Meier recorded one video pointing his Super 8 camera to the south showing this little tree's branches swaying in a reasonably strong wind and good focus. The film camera has a more extended depth of field (DoF) than his Olympus camera which is stuck on infinity focus, so these fir branches are not out of focus in the movie.

If one wonders whether the inhabitants of Fischenthal saw this beamship above the hill towards their East, the answer is, probably not, because we know the craft uses a cloaking system, as explained in the WCUFO section that opens a window of view only towards Meier's camera.

They are Here

Photo #154: 5:21 pm, estimated beamship distance 388 metres.

Figure 132 - Photo #154. From behind the moped. (Source FoM.)

The beamship went back towards the SSE. Meier is looking up the hilltop and to the beamship, now high in the sky above and farther away (Figure 132). The south snowy mountain is again at the left. Are the tree branches at the left from the *eastern woods.*

Meier's moped shows its extension here that pulled a little trailer carrying his cameras and equipment. Neither the large tree nor the little tree is visible. The moped must be very close to the point marked BM on the map in Figure 128.

We can perhaps imagine how difficult it was to ride one-handed to the top of this hill on Meier's simple moped. At times he necessarily used the traction of the motor to pull the trailer while running beside the moped up the steep hill holding it with his one hand and operating the throttle. Stevens mentions this challenging climb in his first book on the Meier contacts *(UFO Contact...A Preliminary...).*

Encounter with a New Beamship

Photo #155: 5:23 pm, estimated beamship distance 384 metres.

Figure 133 - Photo #155. Showing beamship towards the south with Meier's Super 8 film camera on a tripod. (Source FoM.)

The beamship has now moved to the south, towards the south snowy mountain. It is 384 m away in the distance. Here, in this high-resolution copy, we see Meier's film camera mounted inside a frame on a tripod (Figure 133).

Meier probably filmed here what the second film in both the Nippon TV and FIGU Hasenbol documentaries show. The tripod is perhaps to the left of his moped, and the tree branches on the left are probably from the *eastern woods.* Meier made two videos, one from behind the branches of a little fir tree at the *north grove* with his camera sighted towards the south, and the second pointing towards the SW. For his second video, we know Meier recorded at least two films. Unfortunately, no one knows if the other roll, or rolls, of movie film were also lost, stolen, or misappropriated like the two Olympus camera rolls of photo film that never returned from the photo lab.

They are Here

Photo #157: 5:26 pm, estimated beamship distance 76 metres.

Figure 134 - Photo #157. Beamship flying very close yet high in the sky. (Source FoM.)

Three minutes later, the beamship came very close to Meier. Once again, this photo #157 is a full-frame high-resolution copy that anyone can download from the *Future of Mankind* website to zoom and perform image processing and look at the bottom of the craft.

The beamship is located high, at around 45° elevation, so no mountains or trees are visible in this shot. The direct distance to it is 76 m while it is 54 m above the top of the hill and 54 m away from Meier horizontally.

To estimate and plot the beamship's direction on the map, we checked its reflections of the sun. We know the sun at this time is close to SWW in direction, which gives an approximate guide to locate the beamship on the map.

Encounter with a New Beamship

Photo #168: 5:50 pm, estimated beamship distance 60 metres.

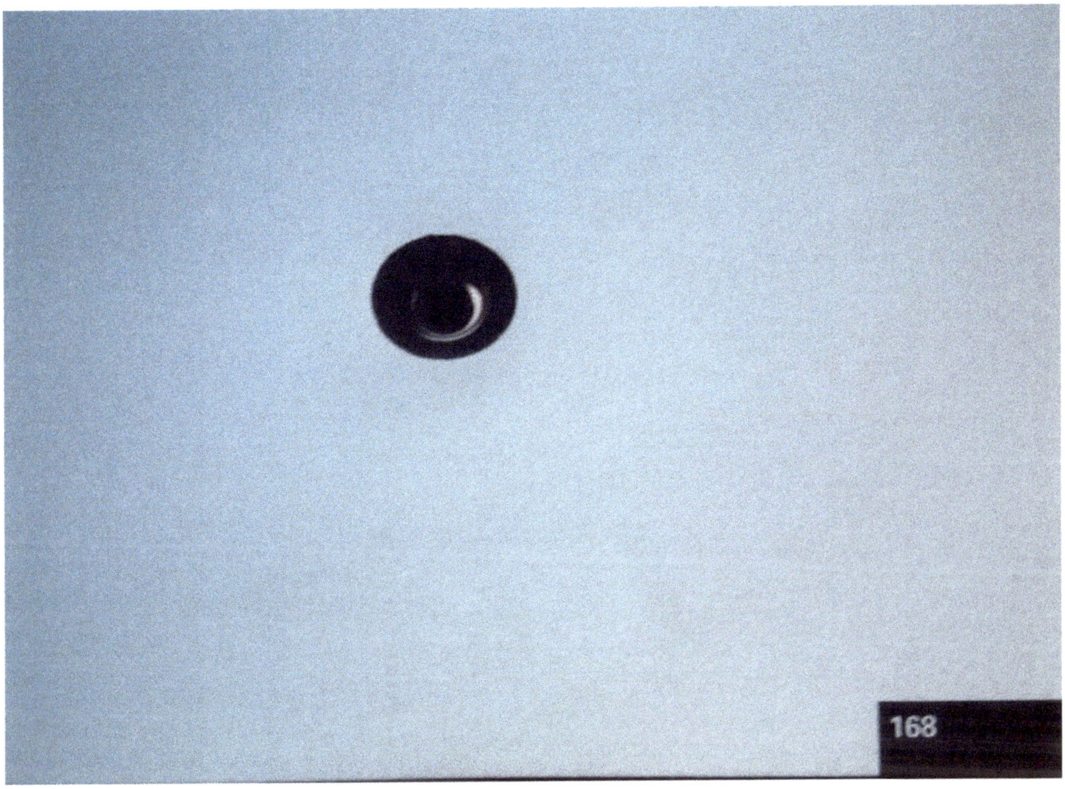

Figure 135 - Photo #168. The beamship again from below (*Photo-Inventarium*, page 85).

Here is another photo (Figure 135) showing the beamship flying above. It may be a cropped image rather than the full-size photo and is a copy from Meier's *Photo-Inventarium* book, page 85.

Its distance away, using the camera formula, is 60 m but it should be longer, around 75 m. As in the previous photo, we can estimate the beamship's direction by noting the reflection of the sun's rays.

The beamship is at an elevation of 55°. For those interested in geometry and trigonometry: to calculate the elevation measure each elliptical axis of the beamship's circular border, that reproduces as an ellipse due to perspective. (Hint: use a sine function).

They are Here

Photo #171: 5:56 pm, estimated beamship distance 75 metres.

Figure 136 - Photo #171. The beamship towards the south. (Source FoM.)

Figure 136 is an extraordinarily beautiful beamship photo taken towards the south snowy mountain (mount Aurüti). Cropped, it gives the mountain more prominence than in the original. Comparing it with other photos, the estimated full size of the frame provides an estimated distance of 75 m.

To get a good shoot, as seen in the FIGU Hasenbol documentary, Meier walked to the west and closer to the *large tree*, so no tree branches from the *eastern woods* appear, or perhaps some appear in the full-size original non-cropped photo.

Recent satellite photos, like those in Google Earth, show more vegetation on this mountain.

Encounter with a New Beamship

Photo #174: 6:02 pm, estimated beamship distance 90 metres.

Figure 137 - Photo #174. The beamship behind the *large tree* and in front of the sun.

Here again, is the remarkable photo #174 with the beamship behind the tree and in front of the sun (Figure 137). Wendelle Stevens described either this photo or photo #164 as "one of the most beautiful UFO pictures ever made" (*UFO Contact...A Preliminary...* page 284). We have to agree. The *large tree* is 50 metres from the camera. Stevens says the *large tree* is "31 feet tall by 21 feet in diameter, and 52 yards away" (*UFO Contact... A Preliminary...* page 351). Our calculations show this tree was 50 m from the camera and 9 to 10 m tall. To take photo #174, Meier was close to the number 1 mark in Figures 125 and 127.

Figure 138 (left) is an enlarged, cropped version of this famous Meier Hasenbol beamship photo #174. The enlargement shows more clearly the atmospheric attenuation, and golden Mie light scatter over the front of the craft.

In *The Pendulum UFO* video section of this book, Part II, we discussed atmospheric attenuation, and how distant objects tend to lose clarity and especially vivid local colour and dark tones. We

They are Here

referred then to this photo taken by Meier at Hasenböl-Langenberg, here called the *Hasenbol beamship*.

Figure 138 - Meier's 29 March 1976 *Hasenbol beamship* photos #174 (left) and #164 (right) with sides and top cropped to show better the atmospheric attenuation and golden Mie light scatter over the front of the beamship.

Encounter with a New Beamship

As the craft's distance from the camera increases, its super carriage and undercarriage tones are losing darkness due to atmospheric attenuation. Its dark undercarriage tone is between that of the nearby grass and tree and the forest on the horizon. The undercarriage darkness is closer to the foreground tones, so we know it is much closer to the tree than the horizon but more distant than the foreground hill. The grass at the picture bottom is uncropped, revealing the full dark tone of subject matter close to the camera. Figure 138 shows a tonal comparison with no added processing; the beamships are straight copies.

Mie light scatter causes parts of the undercarriage to appear even lighter. Still, it is the edges of the craft and the exact centre bottom in photo #164 that give a more accurate estimate of the tone without too much influence of Mie light scatter, although there is still a little scattered sunlight at the craft's extremities.

In the full-photo, the light sky behind the beamship creates the optical illusion that the beamship is darker than it is. Cropping a section of the beamship with its undercarriage and placing it proximal to or over the foreground reveals the differences in tone: the foreground is markedly darker. Figure 139 shows cropped sections of different parts of photo #174 with no changes in brightness, contrast or colour of the sections cropped, just a 200% zoom to obtain a good comparison. Number **3** in Figure 139 is from the beamship undercarriage. The grass nearer the camera is very dark, but the undercarriage is considerably lighter. Even taking the Mie light scatter and sunray dispersion over the craft into count, the difference in tones indicates the beamship is beyond the tree. If it were a model half a metre in diameter or less, it must be at the same distance as the nearby bottom grass in this photo, and its saturation and darkness must correspond to the nearby grass, but it does not. The immense difference between **3** and **6** and even **5** is unmistakable in Figure 139.

In Figure 139, golden Mie light scatter caused by sunlight reflecting off the craft and passing through the corridor of mist between the camera and the beamship tells us the craft is at a reasonable distance from the camera to produce this effect through the light evening mist.

They are Here

Figure 139 - Tone comparison of beamship, foreground, and horizon: **1** the sky, **2** distant mountain, **3** the dark beamship undercarriage, **4** tree trunk and distant mountain, **5** grass near the tree and **6** grass close to the camera.

The abundant Mie light scatter over the front of the beamship creates an outstandingly beautiful photograph and also proves the craft to be quiet a distance from the camera. The beamship is far enough away for an atmospheric corridor long enough to allow sufficient depth of moisture in the light mist to scatter the sunlight

Encounter with a New Beamship

visually as it reflects off the beamship. This effect does not occur within a mere 6- or 7-metre atmospheric corridor of such a light mist. A closer model might give a vivid flash of reflected light, but not this degree of Mie light scatter.

Furthermore, our calculations following shortly concerning this photo and the bird's-eye view plan in Figure 127 show the tree at about 50 metres and the beamship at about 90 metres from Meier.

The Nippon documentary shows the cameraman very close to where Meier shot this photo, and although the Nippon TV camera is a bit further behind it gives a good comparison of the foliage-full tree few years after Meier's photo shooting (Figure144). The FIGU Hasenbol documentary (1998) reveals the stump of the logged *large tree* (inset in Figure 140). The *little tree* is also missing.

Figure 140 - Photo taken during the Meier and Frehner Hasenbol hill visit in 1998. The *large tree* is now cut down (rectangular inset). Also missing is the *little tree* close to it towards the south. We see the *south tree* behind and downhill, which is absent from current satellite photos.

They are Here

It was 1998, during the visit by Billy Meier and Christian Frehner to the Hasenbol hill to record the FIGU Hasenbol documentary, that they first noticed the *large tree* (and the *small tree* and *north tree*) had been cut down. In Figure 140, we see the stump of the logged tree (zoomed detail), and in the documentary, Meier and Frehner film it in detail on the ground. The documentary also shows the top of the *south tree* behind the *large tree*, which is not in satellite images today. Probably these trees were cut down to prepare the area for cultivation. In the same FIGU documentary, Meier says there was another tree that is also missing (the *north tree?*).

Encounter with a New Beamship

Photo #164: 6:05 pm, estimated beamship distance 87 m.

Figure 141 - Photo #164. The beamship is now slightly closer to Meier, behind the *large tree* and again in front of the sun. (Source: High-resolution jpeg from Christian Frehner.)

Photo #164 (Figure 141) is similar to the previous photos #174 and the next covered #175. The sun again is behind the beamship, which is distant as the unusually prominent Mie light scatter reveals. This superior light scatter in photos #164, #174 and #175 is stunning and unseen to this degree in any other photographer's UFO photos that the authors are aware of, and most significantly it proves the craft to be a fair distance away from the camera. Meier is standing at the same place from where he took photo #174. Note, photo #164 is numbered out of shooting sequence which may account for the incorrect time given for it in *Photo-Inventarium*.

Comparing the sizes of the *large tree* in photos #174 and #164 reveals the beamship here to be 3.8% bigger. So a reasonable estimate of its distance is 87 metres. Stevens puts the *large tree* in this photo at a distance of 47.5 metres (52 yards) away from the camera (*UFO Contact... A Preliminary...* page 351; Yaoi / Elders *UFO...Contact From The Pleiades* page 37). We found it is more likely to be 50 metres. If we assume these figures correct, the beamship is 37 metres or just about 40 metres behind the tree. This photo #164 graces the dust cover of Stevens *UFO Contact from the Pleiades: A Preliminary Investigative Report*.

A close inspection of the exceptionally bright light scattering pattern on this photo reveals it to be the photo Stevens (or his team) scientifically analysed in the 1970s. The photo provides compelling evidence of a genuine shot of an unconventional flying object or beamship. Stevens made an internegative of this photo to create a high-quality print, which conclusively shows the tree over the craft, putting the UFO at a considerable distance from the camera. Below are Stevens' comments regarding the analysis (Figure 142):

> This [inter]negative was laser scanned by America Color Corporation of Phoenix, Arizona, using an identical model HELL Chromograph DC300 scanning computer with an Argon laser beam....
>
> American Color made 4 separate color separation negatives by the same method and we printed them back in 4 color lithograph to obtain the result you see here [Figure 142] This color print was then photographed to get this picture.
>
> Here it may be clearly seen that the symmetry on the left side of the rim of the spacecraft is broken by a forked branch of the tree that is BETWEEN the ship and the camera, clearly positioning the ship behind the tree as stated by the witness (Stevens *UFO Contact... A Preliminary...* page 352).

The authors do not have access to a copy of Stevens' high quality original computer-generated print. They do not know anybody that does, but zooming the Frehner image available of photo #164 does show the beamship behind the tree (Figure 143). Besides, the Mie light scatter in this light evening mist dictates a reasonable distance to the object. (Readers can zoom in on the High-definition jpeg in the FoM photo gallery.)

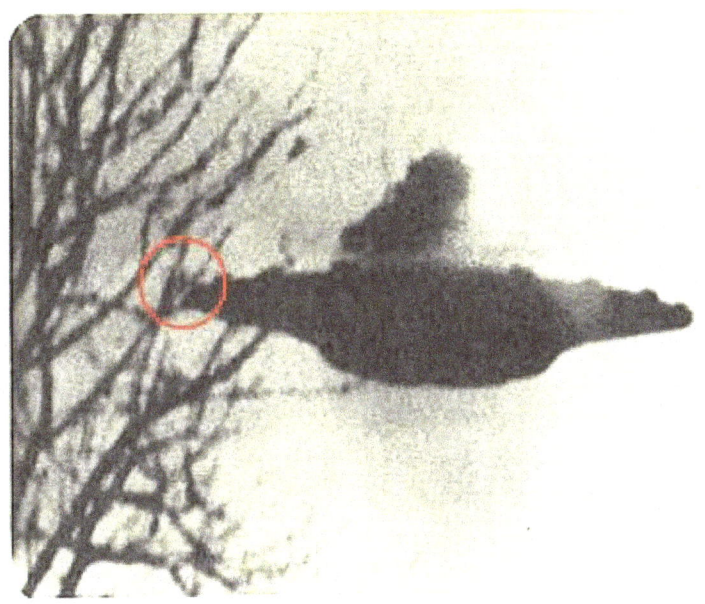

29 March 1976, Hasenbol Langenberg, 19.50 This image was produced from the same internegative as used to print the previous photograph. This negative was laser scanned by American Color Corporation of Phoenix, Arizona, using an identical model HELL Chromograph DC 300 scanning computer with an Argon laser beam, using the same process as Interrepro, A. G. in Basel

American Color made 4 separate color separation negatives by the same method and we printed them back in 4 color lithograph to obtain the result you see here. This color print was then photographed to get this picture.

Here it may be clearly seen that the symmetry on the left side of the rim of the spacecraft is broken by a forked branch of the tree that is BETWEEN the ship and the camera, clearly positioning the ship behind the tree as stated by the witness. When we separated the tree from the ship by color contouring in a computer the evidence is considerably more impressive. More of this could be done if the process were not so expensive.

Figure 142 - Top: An early scientific-analytical report on photo #164 proves a forked tree branch is between the ship and the camera, positioning the ship behind the tree (Stevens *UFO Contact... A Preliminary...* page 352).

They are Here

Figure 143 shows our zoom and enhancements of photo #164 with the tree branches visibly over the left edge of the beamship. The top right picture in Figure 143 is from FoM. Bottom images are Villate's two Photoshop details of Frehner's photo #164. [13] The bottom right picture shows Villate's traced branches over minor details shown in the bottom left image.

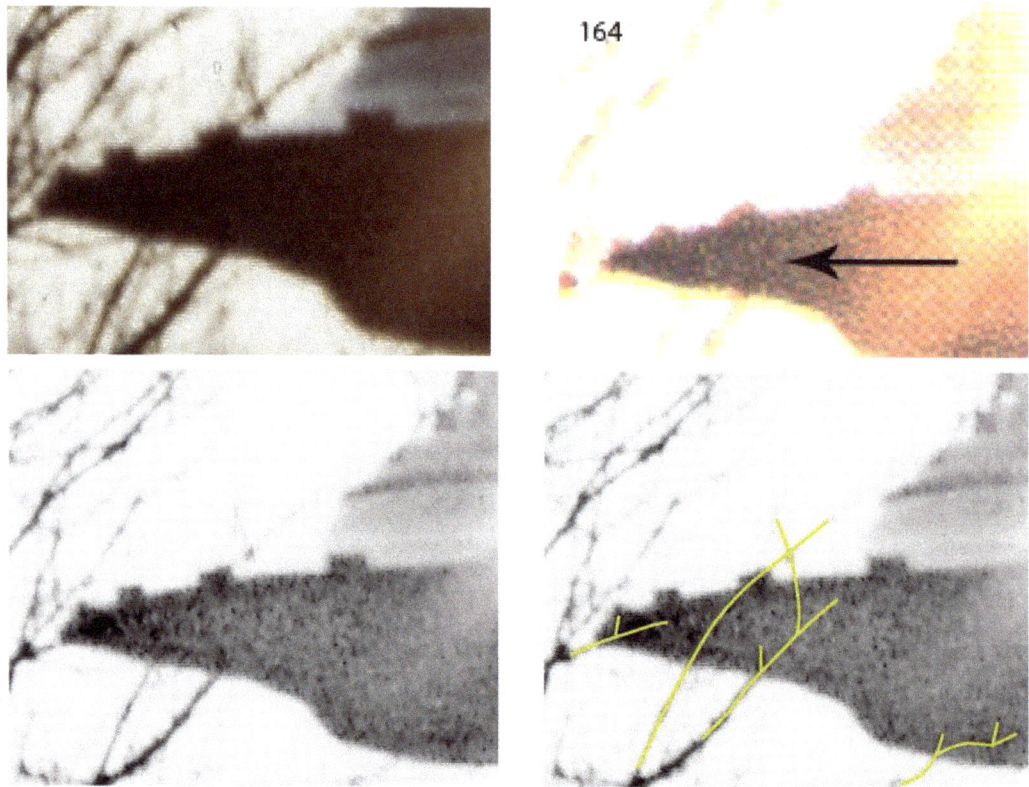

Figure 143 - Top left: Zoom in of Christian Frehner's jpeg of photo #164. Top right: FoM picture. Bottom: Villate's two Photoshop details of Frehner's photo #164. [13] Right: Branches traced over minor details shown in the left picture.

Photo #175: 6:04 pm, estimated beamship distance 88 m.

Figure 144 - Photo #175. Top: Old Wendelle Stevens photo and an enlargement showing the beamship at a 5% clockwise dip. (Sources FoM.) Bottom: Left, Yaoi / Elders *UFO...Contact from The Pleiades* Japanese version page 44). Right, Tom Welch in 1978 standing by the *large tree* in full leaf with two years further growth. Note the tree branches again over the edge of the craft.

They are Here

Photo # 175 (Figure 144) shows the beamship with a 5% clockwise dip and again with tree branches in front of the beamship. Figure 144 bottom is the authors' scanned image from the Junichi Yaoi Japanese version (page 44) of the Britt and Lee Elders *UFO...Contact From The Pleiades Volume I*. Interestingly, it shows Tom Welch in 1978 standing by the *large tree* in full leaf with two years further growth. It confirms just how tall and broad this tree was and that Stevens figures of 31 feet tall by 21 feet in diameter are accurate.

The authors were unable to locate a high-quality print of photo #175, but with tools like Gimp or Photoshop, the beamship is seen once again behind the tree (Figure 145).

Figure 145 - Analysis by Sean Gibbons, H.Dip. Digital Media. Meier's #175 bump-mapped on image manipulation program 'Gimp' (Source Deardorff website). Note: Image tilted 10% anticlockwise compared with the original (see Figure 144).

Encounter with a New Beamship

Photo #176: 6:09 pm, estimated beamship distance 59 m.

Figure 146 - Photo #176. The beamship now moves closer, back to the SW (Source FoM).

This full-frame beamship photo is again towards the SSW. It has now moved closer, and the following photos reveal more exciting details of this flying machine.

Being a high-resolution *wallpaper* size image, a zoom-in of Figure 146 reveals the "ports" on the rim of the ship, and we estimate there are 20 of them (Figure #147). Also, we see windows, numbering around six, which appear to be orange. The beamship undercarriage is very dark while its upper body is light blue reflecting the sky.

The top section is also noticeably featured with several parts rather challenging to see in detail. There could be several small spheres. The top section is narrower in one direction than the other. Here, in Figure 147, we see the upper body from the narrow side. Right on its top is a ring cited by critics as suggesting it was a hung model, albeit an extraordinarily complex one. Others have suggested it is an attempt to discredit Meier by people who

They are Here

tampered with the photo. Yet others say it is a natural feature. Notably, it is not in every photo of the series. Intriguingly, as Devine mentioned to us, it is also similar to the ring on top of the Al Aksa mosque/Dome of the Rock (Wikipedia).

Figure 147 - Zoom-in of high-resolution photo #176. The beamship has an estimated 20 dark rectangular ports around its edge, and six windows, two visible here. A "ring" sits on its top. (Source FoM.)

The top ring could be an antenna, but it also reinforces the sceptic barrier and contributes to the worldwide UFO controversy. It is an ideal detail for critics to use as an exit door, although not one among them has suggested in over 45 years precisely how to make this "model." Critics claim Meier made models, and here we see a ring which they claim Meier used to attach a nylon cord from which he supposedly hung the model. In a section ahead, however, compelling details show why even with this single little feature, this cannot be a model.

The top or upper body of this beamship sometimes looks wide as in photo #179 (Figure 148) and at other times narrow as here in photo #176. When it looks wide, the ring is viewed from the edge and appears like a single vertical pole.

Photo #179: 6:09 pm, estimated beamship distance 55 m.

Figure 148 - Photo #179. The beamship moves closer to the camera and rotates on its vertical axis now showing the top section from its wide side. (Source FoM.)

Like the previous photo, photo #179 is a full-frame high-resolution *wallpaper* copy. Again, we see intriguing details of the ship that has moved four metres closer to the camera. In a zoomed image (Figure 149), what looks like small spheres are visible above the windows on the top section.

In one of the two videos available, we notice the beamship rotate. It might have turned 90 degrees on its vertical axis to reveal the top section from its wide side, and to show the top ring antenna on edge (see Figure 149).

Checking the location of the distant mountains and comparing them with the previous photo, the direction is about the same, towards the SSW.

They are Here

Figure 149 - Zoomed details of photo #179. Notice the dark lines at the base contours. (Source FoM.)

The ring antenna, now from the edge, looks like a single vertical pole. Billy Meier explained that this ship could accommodate three crew members, and according to Wendelle Stevens, Semjase and Quetzal are said to have come at this time, and Quetzal exited the beamship and spoke with Meier:

> One of the aliens, Quetzel [sic], descended on a beam of light and stood with Meier near the tree where they talked for some minutes. Then Quetzel [sic] returned to his ship...(*UFO Contact...A Preliminary...* page 284).

Studying the image in detail, a dark contour, a cartoon-like line, surrounds the bottom of the ship. The effect is common in film emulsion photos. A bright area like a radiant sky adjacent to or close to a dark area, like the beamship undercarriage, produces a photochemical reaction that creates this line. Jim Diletosso, an early investigator, found how this effect changed, showing different widths of these lines depending on whether the UFO is a significant object or a smaller scale model. See the documentary *Contact – 'Billy' Eduard A. Meier Documentary by Wendelle Stevens* (1982). Diletosso concluded that the UFO photos from Meier show big objects, compatible with Meier's statements of seven-metre diameter beamships.

Encounter with a New Beamship

Photo #181: 6:20 pm, estimated beamship distance 50 m.

Figure 150 - Photo #181. Full-frame high-resolution beamship photo. (Source FoM.)

In this final full-frame high-resolution photo #181 (Figure 150) the beamship has moved towards the south, and a few tiny tree branches, probably from the *eastern woods,* enter the picture frame at the bottom left corner.

The ship is very close to Meier and at around 30° in elevation. As in the previous photo, we see the top section from the wide side and the ringed antenna on top looks again like a single vertical pole.

Noticeably, the beamship here is unusually asymmetrical, with its left side longer than its right, caused almost certainly by a pincushion distortion effect of the camera lens which occurs with objects close to the photo edges and corners. The same distortion is in some WUFO photos taken in Meier's courtyard in front of his house.

They are Here

Available videos

There are two existing videos of this beamship's flight demonstration. Meier shot Video 1 with his Super 8 camera behind a little fir tree in the *north grove* with the south snowy mountain as a backdrop. Video 2 looks towards Fischenthal in the south-west. For the second video, Meier used two film rolls. The transition from one film roll to the other with a marked frame is seen in Figure 85, in Part II. Both videos show characteristics of the beamship flights that cannot be explained by the use of a small model hung from a cord on a rod. The next chapter discusses this in detail.

Figure 151 - Video 1 screenshot. The beamship is towards the south with the camera behind a fir tree in the *north grove*.

These videos are part of the Nippon documentary and the FIGU documentary *Demonstrationsflüge / Demonstration flights (IFOs, not UFOs)*: Video 1 at 32:40, and Video 2 at 35:44. They also appear in the documentary *Billy Meier: The Hasenböl/Hasenboel Footage*.

Encounter with a New Beamship

Figure 152 - Video 2 screenshot. Now the beamship is towards the south-west. Meier zooms with the camera several times.

In Video 1, the beamship moves left to right then right to left (east to west and back to east). The movement looks like a forced half pendulum, stopping abruptly sometimes remaining completely stationary and repeating the movement left to right. Branches of a little fir or Norway Spruce are in focus right in front of the camera, which is pointing south.

Video 2, with the camera installed somewhere else facing south-west, shows the beamship almost static with a soft movement to its sides. The beamship rotates on its vertical axis showing a wobbling motion.

In the video, a few surprising flashes of light from the beamship have been "explained" as sun reflections on flat surfaces in the ship's rim ports, and flat surfaces on the upper superstructure. These flashes, however, are shown and discussed in the following section.

Meier uses the camera zoom several times, enlarging and reducing the beamship size as it fills more or less of the picture during the recording.

No Trick Here

In this chapter, we evaluate the possibility that Billy Meier used a small model or trick photography. Some sceptics claim Meier's fabricated his photos with small models hung on cords and positioned close to small trees, or used a false perspective trick. Part I and Part II of this book demonstrated how the WCUFO and *The Pendulum UFO* respectively provide compelling evidence that Meier's UFOs are sizeable beamships, and cannot be tiny models. Here, we further verify whether a model was used to produce The Hasenbol Beamship Demonstration.

Behind the tree

Abundant evidence proves the *large tree* in front of the beamship existed. Before being logged, the tree appears in photos in Wendelle Steven's book, the Yaoi / Elders *UFO...Contact From The Pleiades* book (Figure 144), and Yaoi's Nippon TV documentary (Figures 153 ~ 155). We also find this tree and the other missing ones in the aerial photo from 1972 (Figure 125, top).

Figure 153 - Comparison of photo #174 (left) and a snapshot from the Nippon TV film (right). Meier and Yaoi are to the right of the *large tree*. The camera is equidistant from them and the *large tree*.

They are Here

We know Meier did not use a little tree here because abundant photos and films show in bright daylight that it was indeed a big tree standing 31 feet tall. That it has since been logged, as seen in the 1998 FIGU Hasenbol documentary, is an irrelevant distraction. We know for a documented fact that it still existed in 1978. See Tom Welch next to the tree in 1978 in Figure 144, and Junichi Yaoi next to it a few years after the demonstration in Figure 153 (right). Moreover, the *large tree* stump remains, at least to 1998.

In the Nippon TV documentary, Meier and Yaoi are on the hill location with the tree. Meier is demonstrating how he mounted his Super 8 film camera on his tripod while the documentary cameraman is very close to where Meier was taking his three beamship photos of the craft in front of the sun. The distant mountains in Figure 153 share the same orientation in both images, although slightly lower in the Nippon TV documentary. Meier was either a bit closer to the tree or his camera a bit higher, making for a better view of the background mountains. So we know the two trees in Figure 153 are the same tree.

Observing the Google Earth images shows the terrain slightly ascending and then descending toward the *large tree* (Figure 127-top), and beyond it is an abrupt descent where the *downhill tree* stood.

When Meier and Yaoi stop by the *large tree*, we notice just how big it is (Figure 153). The tree, judging by their height is 11 metres tall or a bit over 30 feet; precisely what Stevens found. Maybe during the photo session on 29 March 1976 a couple of years earlier it was a little smaller, at around 10 m.

Figure 154 shows the width of the beamship. If it were precisely at the same distance as the *large tree,* its size would have been four metres. At 90 metres from the camera, its size might have been seven metres. Meier remembers the beamship being closer to the tree, and maybe he took pictures of it there, but if so those photos were either lost or stolen.

Interestingly, Yaoi has another YouTube video in which he interviews research scientist Marcel Vogel from IBM. Vogel explains from the 38-minute mark in the film *Beamship – The Metal* how we could not reproduce this beamship metal on Earth at that time in the 1980s, and that it shows clear evidence of cold fusion of the various metals, including rare-earth metal.

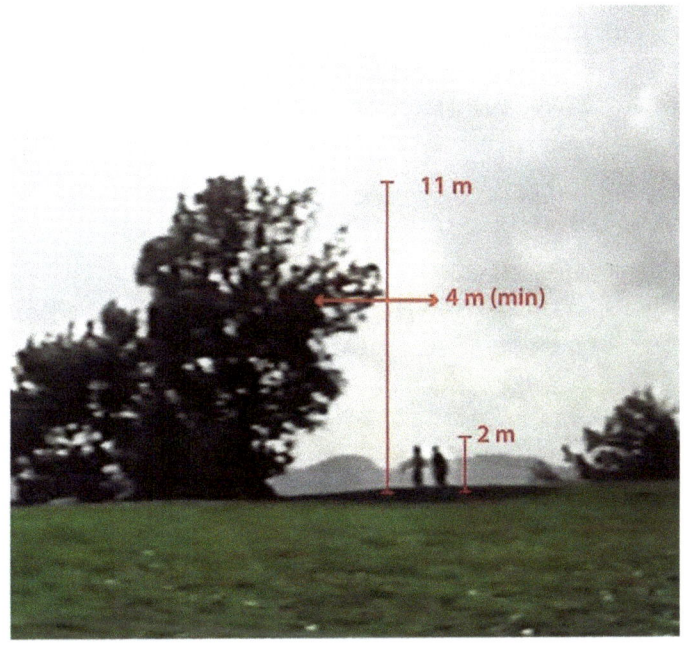

Figure 154 - *Large tree* height estimation and beamship size had it been precisely the same distance from the camera as the tree.

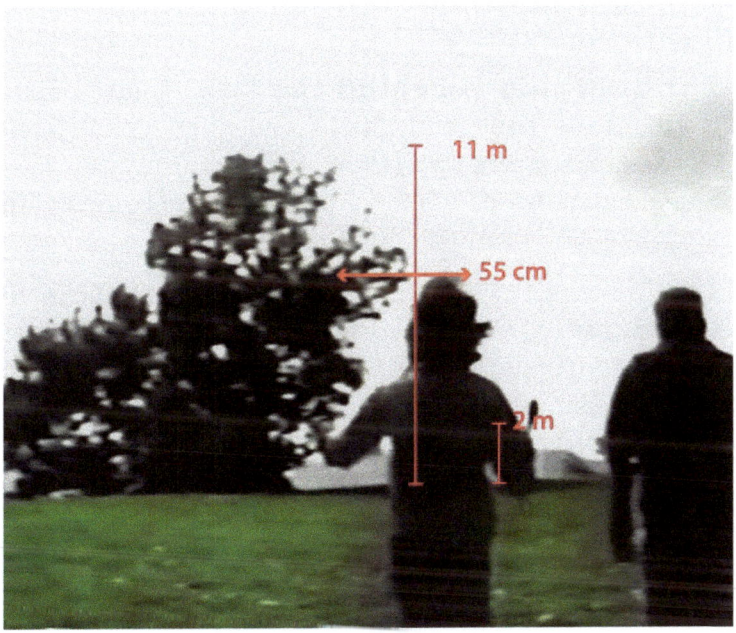

Figure 155 - Had the beamship been a small 55 cm model, it would be where Yaoi is standing. (Meier is on the right).

Figure 155 shows the model width. If this beamship were a 55 cm model (the usual size sceptics guess), it would have been just six metres from the camera. When Yaoi and Meier start to walk towards the *large tree* Yaoi is six metres from the camera. So he can be used as a measure for a hypothetical model. Notice how near Yaoi and Meier are.

The key to knowing if the beamship is behind the *large tree* is in studying the photographic and other evidence. Previously we have shown the tree is in front of the beamship, in Figures 142 and 143, for the photos #175 and #174. The image was enhanced with the free computer tool Gimp, and Photoshop that shows tree branches between the beamship and the camera, placing the beamship behind the tree. So the beamship is more than four metres wide.

Another factor in the photo leading to the inevitable conclusion that this beamship cannot be a small model close to the camera is its distant atmospheric attenuation. Mie light scatter over the front of the beamship was previously discussed under Figures 138 and 139, again, proving the object is a considerable distance from the camera.

In summary, we conclude:

- **The beamship is behind the tree.** A few branches cross in front of the beamship. See Figure 145 and James Deardorff's website article with an image and analysis by Sean Gibbons of Meier's picture #175. Also Figures 143 and 144 with similar findings.

- **The beamship is well over seven metres away from the camera and probably at about 90 metres**. Mie light scattering around the UFO, or beamship (photo #174), dictates a long corridor of a misty atmosphere between the camera and the UFO. There is not enough water vapour in the atmosphere over a distance of six or seven metres to create such an effect.

- **The beamship might have been seven metres in diameter.** If so, calculations put it at about 90 metres away from the camera.

No Tricks Here

The "impossible fishing rod"

Part II of this book discussed the different possibilities of using a model hung from a cord. One was using a fishing rod, and we mentioned the *Ridiculously Long Fishing Rod* of five metres. However, to create The Hasenbol Beamship Demonstration, an impossible 1975 fishing rod of 15 metres in length would be necessary, which is, of course, impossible.

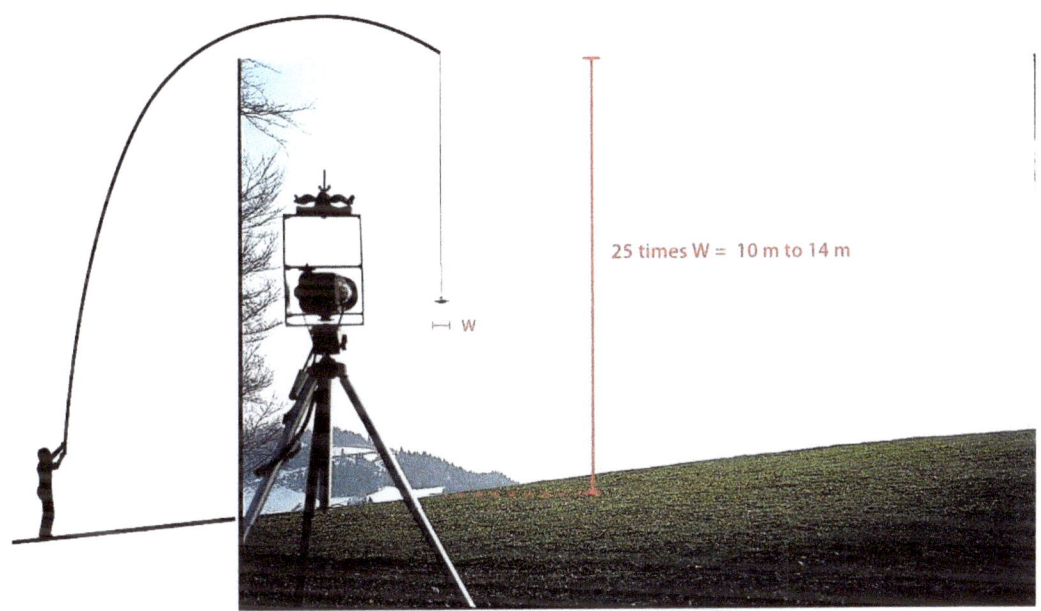

Figure 156 - Photo #155 with the estimated necessary fishing rod length of over 15 metres. The human figure is drawn to scale.

Some photos of the distant beamship show a vast space from the ground to the top of the image. Figure 156 shows such a picture, #155, and in Video 1, it is noteworthy that Meier must use a fishing rod longer than 15 metres.

Because of such intricate details in the photos and videos, any model would have to be 40 ~ 55 centimetres in diameter. A tiny model of five centimetres in diameter is too small to make with such intricate details. If Meier used a model in photo #155, he had a vast distance from the ground to the top node of the fishing rod: a minimum length of 25 times the model diameter. In this photo, that must be 10 ~ 14 metres. See Figure 156 with a human figure drawn to scale for a 40 ~ 55 cm model.

It may not be impossible to construct a suitable lengthy supporting system. Still, it is difficult to imagine a one-armed man, driving a moped up steep hill terrain, carrying such an unwieldy rod to fake a real beamship when collapsible fishing rods of this length were indeed unavailable. Moreover, from our experiments, we know that handling such a long fishing rod would require more than one person to hold and control it. The rod's weight plus the model's weight, even a small model, would require maintaining a force of more than 30 kilograms to prevent the rod falling; and how would he prevent it from breaking? Why would Meier not do it more efficiently in a more accessible place and hang the model from a tree? Handling the required model with such an impossible rod is extraordinarily complicated, even for a group of people. There are no available trees at the location for practical support to assist in making this a photo trick.

Moreover, watching the Video 1 clip with the beamship moving left to right with the south snowy mountain in the background, immediately impresses on the viewer how difficult it would be to replicate this with a UFO model using an *impossible fishing rod*. More details ahead.

Beamship rim flashes of light

An exciting feature of Video 2 is its full flashes of light on the beamship's right side edge, bottom section, where we find 20 ports, and more on the beamship's upper section. A snapshot from this video, Figure 157, shows two such flashes.

What causes these flashes? A simple explanation could be sunlight reflections. Noticeably, in Video 2, the beamship rotates, giving flashes of light always coming from its right side facing the sun in the west.

If the craft were a model, it would undoubtedly already be a very complex one to make, requiring many intricate details. Yet now it needs the further ability to rotate in a very controlled fashion and the addition of several new small reflecting surfaces at certain places. Judging by the eye, the flashes could be either just reflections of sunlight or a confounding effect caused by the beamship. Stevens, detailed scientific analysis in his *Preliminary Investigation Report*, however, concluded the flashes are produced by the craft projecting beams of "bright coherent white light":

> The analysts examining the movie footage from this last event were amazed to find that what at first looked like a flash of reflected sunlight from a part of the rim of the ship and an area of the dome, was in fact a projected beam of bright coherent white light from something. The beam is clearly seen and it is sharply distinct and does not spread out as it leaves the ship (http://theyfly.com/PDF/PhotoAnalysis.pdf page 12).

Figure 157 - Video 2 flashes of light on the beamship's right side edges.

Unexplained movements

In both videos, beamship movements are hard to explain. We made and present in Figure 158, a composition of over a dozen beamship positions from Video 1.

The movement is in a long arc, and if it were a model hung from a cord, this pendulum node (where the cord connects with the fishing rod), must be very high, several times the distance from the model to the top of the photo. Even the 15-metre-long *impossible fishing rod* would be too short; a device two or three times longer would be required. The node, noticeably, would need to be located very high at the centre of the arc described by the beamship, (indicated as a dashed vertical line in Figure 158). So we have half of the pendulum movement, the left side, but not the right side. It means the beamship never moves beyond the middle point in the pendulum trajectory, not beyond the vertical line in the figure. The beamship always stops at the midpoint, or lower point in the arc it describes. It could work with another assistant at the left side standing at around 10 metres above ground level, pulling a new cord connected to the model to stop the movement before it goes beyond the midpoint. But now a cumbersome, long ladder is needed to access the required equipment for the demonstration to allow this additional assistant to pull the model to a stop.

Noticeably, the UFO stops abruptly, in the lower part of the arc, and restarts the movement towards the left (The assistant on the top of the ladder must pull the cord very quickly to help the model go to the left).

Also, in Figure 158, the beamship is sometimes above or below the expected arc of a pendulum movement (see the arc in Figure 158). If the UFO is a model, the fishing rod's flexibility could explain it. In this case, we can also estimate the cord length if we check the pendulum period as we did in Part II of this book. Here we see just a fraction of the pendulum movement; actually, a quarter of a period.

To estimate the cord's length by using the pendulum function, we must first measure the left half of the semi-period that the UFO travels, from the left-most position to the middle point where the vertical line and the arc intersect in Figure 158. The period is then four times this half of the semi-period travelled. From the period we can then calculate the cord length as done previously in Part II. However, the cord length, when calculated, changes dramatically, from 46 metres up to 137 metres, with an average of 80 metres. How can this possibly work?

No Tricks Here

Figure 158 - Composition of over a dozen UFO positions in Video 2. The UFO is moving left to the right (east to west).

For a model, presumably, an assistant on the left pulls it to its left-most position. Whereon, the cord is released, and the model falls freely like a pendulum, passing to the middle point, where it is stopped by the same assistant pulling the cord again. But how can this create the observed sizable variation in the pendulum length? Even with an 80-metre-long cord, we now need an extremely long fishing rod of perhaps 90 metres. Such a rod does not exist.

Furthermore, correlating the UFO model with the arc described in the film makes the model about 55 centimetres in diameter. It is outrageously complicated to attempt hanging a 55 cm model from a 90 metre elevation, with an assistant on the left side holding an excessively long cord located at the same height on the top of an exposed and windy hill. Again a UFO model does not fit the data.

Even this first look at the craft's unexplained movements exposes how difficult it would be to perform the beamship demonstration with a model hung from a cord. But other details make it so complicated as to be perhaps impossible. Here are four.

The wind does not affect the model

In Video 1, we see branches from a small tree, possibly a fir or a Norway spruce (Figure 158) that sway considerably, indicating a strong wind. At the same time, however, the model stops and remains static in the sky. Assuming a model, why is the strong wind on the exposed hilltop moving the branches but not the model? The model must move correspondingly.

Could somebody be shaking the branches? Perhaps it is required to compare the movement caused by the wind with that caused by manually shaking the branches. If somebody moves them, the demonstration now needs another assistant. It is improbable there was not even a soft breeze on the top of this hill. Witnesses have said it is usually windy there. Any exposed hilltop wind as in Meier's film would move the model, but it is sometimes wholly static.

The model stops abruptly

In Video 1, we notice how the beamship sometimes stops very abruptly, which is not possible using a fishing rod. As previously pointed out, using a model would require a horizontal cord and an assistant must pull it to stop the model.

The beamship flips before it moves

Video 1 shows the same effect found in *The Pendulum UFO*: the beamship flips too soon. When moving a cord-hung model from one side, the normal behaviour is a gradual rotation that starts after the execution of the horizontal movement. See section 5, "The sudden UFO flip" in Part II of this book.

In Video 1, the model flips and then moves, and it sometimes makes a small jump before moving. In a standard model hung from a cord, a flip or jump would not be required, just the effect of gravity to move the object. Nippon TV found in their analysis that the UFO looks brighter when the movement starts. The bright flare observed can be caused by a sudden flip of the beamship, causing it to face and reflect the sun's rays. This behaviour of flipping too soon cannot be explained in a standard model, not even with new cords and participants.

Wobbling

One of the most intriguing aspects of this beamship is its wobbling. It is not the usual wobbling of a cord-hung model. We found this effect in *The Pendulum UFO*. (Refer to item 6, "UFO Wobbling", in Part II of this book).

Figure 159 - The wobbling centres at a point close to the base of the beamship. A hanging model must flip or wobble centred close to the cord contact point or in the model's centre, not its base.

Here, however, it is more evident. A hanging model must flip or wobble centred close to the cord contact point or the model's centre, not the base (see Figure 159). The video flips noticeably happen around a point very close to the beamship's base, making the wobbling closer to that of the spinning top children play with than the wobbling from a cord-hung model.

They are Here

The big question is: Why is it wobbling in this way? Sometimes it is entirely stationary and yet it wobbles. Why? If it were a hanging model with the wind causing it, or if somebody was pulling down the edge of the model using another cord, why does the model not start swinging side to side like a pendulum?

Conclusion: no trick or model

- Meier was on a hilltop, fully exposed to the surroundings. With a house and little road just a couple of hundred metres away, several witnesses could readily spot him and any assistants holding a 10-metre ladder and a 15-metre-long ad hoc rod – and neighbours do occasionally watch Meier. A secret trickster is most unlikely to choose such an overly exposed area to perform a secret trick with cumbersome equipment that would invite spectators.

- At least two photos (#164 and #175) and probably a third (photo #174) show the beamship behind the large-tree, proving the craft to be a large object of well over four metres and probably around seven metres in diameter. However, a couple of tree branches in #164 read ambiguously, as either in front of or behind the tree. If they are behind, the ship lies between a couple of the tree branches and so is at precisely the same distance from the camera as the tree, making it four metres in diameter instead of seven. To definitively establish which size requires a more detailed analysis, which so far has not proven possible for us.

- If this were a small model, it would have to be very close to the camera, making it totally out of focus because the camera is always focused on infinity. Yet the photos show the object always in full focus.

- Due to atmospheric attenuation, the beamship, especially its dark undercarriage, is losing tone due to its significant distance from the camera.

- There would be no Mie light scatter over the craft if it were a model within a couple of metres from the camera. Mie light scatter present over the craft eliminates the possibility of a small model close to the camera.
- Unusual beamship movements like the wobbling, flipping too soon, or its half-pendulum movement (not full pendulum trajectory, left to right) remain unexplained with the little model theory.
- The Hasenbol Beamship Demonstration cannot be explained by saying Meier used a scale model hung from a cord. Such an "explanation" does not fit the facts.

All the evidence shows no tricks here: but instead, a significant and extraordinary unconventional flying object performing remarkably on an exposed hilltop one bright and windy March evening in 1976, an event now known as The Hasenbol Beamship Demonstration.

Annexes

Annexe A

Billy Meier's Olympus, Ricoh, & movie camera

The Olympus 35 ECR:

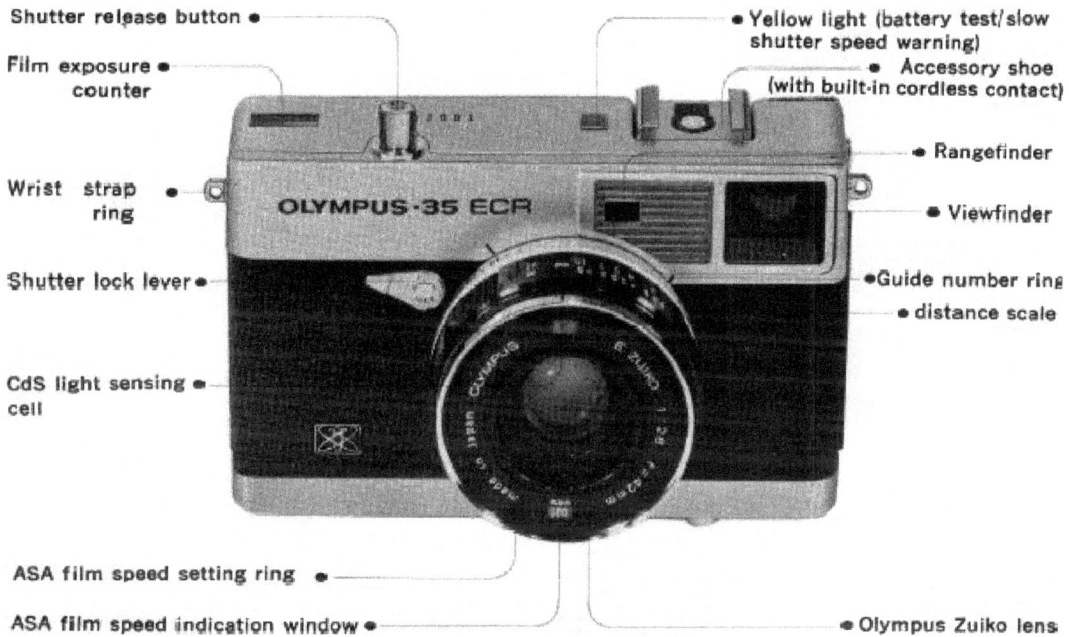

Figure 160 - Olympus 35 ECR parts.

Specifications for the Olympus 35 ECR are in the manual, page 2 available 24 August 2019 free from
https://www.manualslib.com/manual/771811/ Olympus-35ecr.html?page=2#manual.

Also, access: http://www.manualslib.com/products/Olympus-35-Ecr-3052477.html.

Alternatively, access: www.orphancameras.com.

They are Here

The Ricoh SLR 55mm:

Figure 161 - Ricoh SLR camera.

For a free PDF of the Ricoh SLR manual with specifications, click here: https://www.butkus.org/chinon/ricoh/ricoh_singlex_tls/ricoh_singlex_tls.pdf. 24 Aug. 2019.

Alternatively: http://www.butkus.org/chinon/ricoh/ricoh_singlex_tls/ricoh_singlex_tls.htm.

Also, access www.orphancameras.com.

Annexe A

Nalcom FTL 1000 Synchro Zoom:

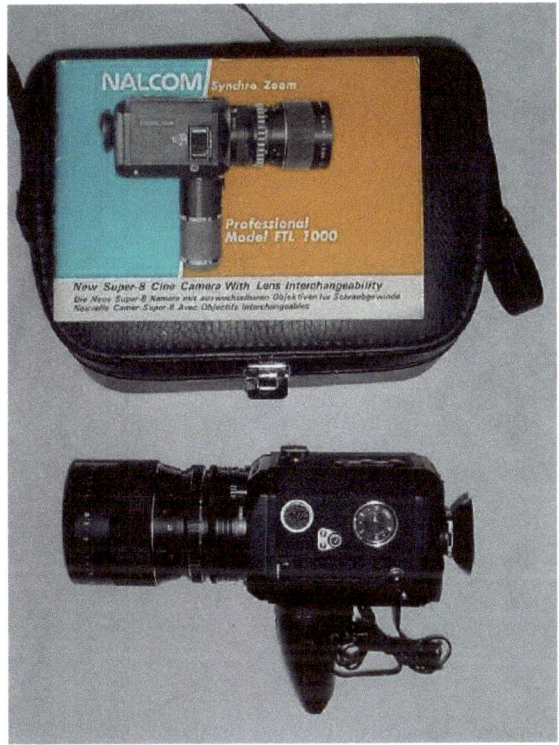

Figure 162 - Nalcom FTL 1000 Synchro Zoom that Meier used.

- Nalcom FTL 1000 Synchro Zoom
- Year: 1973-76. Original price in England (year of introduction): £176)
- Weight: 1573 g
- Lens: Shinkor Zoom 1,8 / 6,5 - 65 mm (removable lens)
- Split Image Focusing
- Auto / Manual Zoom with 2 Speeds
- Frame rates: 18, 24, 36 + single frame
- Manual / Auto Exposure
- Flash contact
- Fades
- Remote control socket. No sound. Made in Japan
- Instruction manual online:

They are Here

http://www.mondofoto.com/manuals/nalcom+ftl+synchro+zoom+1000+pro/

http://sper8wiki.com/idex.php/Nacom_FTL_1000_Synchro_Zoom.

Annexe B

Disclosure Project witness report

This Disclosure Project witness report was received directly by the authors from Dyson Devine in a personal email dated 24 January 2020.

Sergeant (Ret.) Dyson A. Devine's United States Air Force witness testimony as provided to the Disclosure Project

I am prepared to swear under oath that what I state here is true to the best of my recollection.

After some background context, I relate what I was told the unofficial procedure was by the USAF when dealing with radio recordings pertaining to UFO encounters, and the consequences of defiance of these procedures by personnel involved.

I enlisted in the United States Air Force in 1968, and after about a year of training in the States, I was employed as an air traffic control radar technician in Europe until I was relieved of active duty in 1972. I elected not to return to the U.S. and migrated to Australia. After my required two subsequent years as a "ready reservist" I was honourably discharged in 1974. I became a naturalised Australian citizen in 1977.

I served for about three years with the Air Force Communications Service (AFCS) in Detachment 9 of the 2063rd Communications Squadron. I was stationed at Echterdingen Airfield, which was on the other side of the single runway that we shared with Stuttgart civil airport. I was a sergeant.

They are Here

It was my responsibility to maintain and repair an MPN-13 search and precision Ground Controlled Approach (G.C.A.) radar unit positioned very close to the edge of the runway, at least a mile from our barracks. The "M" designates that we were a mobile unit, consisting of a main operating van and a back up van (for our spare sub-units) that were ganged together as wheeled units that could be towed. We provided support for the Stuttgart based U.S. Army Headquarters in Europe, but our ancient radar unit was already technically redundant due to the introduction of Instrument Landing Systems (I.L.S), so we had very little military traffic and we primarily served as an aid to training and as an emergency resource for the occasional light civilian aircraft lost in the fog. My two to four fellow specialists and I took great pride in our work and were formally recognised for the very high technical quality of our old unit.

In all but the first part of my tour of duty there I worked as the only rostered on repairman, doing a 24-hour straight shift from 0800 to 0800. I usually had the company of at least two radar operators who initially worked days only, but eventually were rostered swing shifts until 2300 hrs. For the most part the atmosphere was relaxed and friendly, with plenty of time for bored career operators to tell their war stories. Apart from the aforementioned units, we had another trailer which we repairmen used as a maintenance van, and there was a fourth trailer, known as the creature comfort van, which is where the operators and occasional ground power crew stayed when they were not required elsewhere.

I did not believe in flying saucers, and attributed the regular appearance of large, crisp, single sweep returns on my search scope as inexplicable random intermittent electronic artefacts of the hardware, which we repairmen naturally had no desire to call anyone's attention to. We never spoke about it, even to each other. One other reason that I could not imagine that a real object could be providing such a return, aside from the fact that I never remember seeing more than one at a time (one sweep of the radar beam) was that they were invariably very large and crisp. Sometimes a big aircraft can show a slightly asymmetrical blip, caused by - for instance - a large tail assembly, but these were always wonderfully uniform, as if

Annexe B

reflecting off a large radially symmetrical metal object. I never tried to establish a pattern, but with the benefit of hindsight, I believe their occurrence became more common as the war escalated and our airfield became much busier.

One particularly quiet evening, apropos of nothing much in particular, I innocently asked the three on duty radar operators, "What is all this stuff about UFOs anyway?" I was at least as surprised by the reaction I got, as they appeared to be by my question. The atmosphere instantly turned icy, and nobody moved a muscle except long-serving Master Sergeant Graham, who simply slowly and wordlessly got to his feet and left the van without looking at me to go out and sit in the radar unit. After some more gentle prodding from me, Technical Sgt. Green started to tell me about the de facto standardised USAF procedure for dealing with this matter. The other operator, whose name escapes me, initially objected to Sgt. Green telling me anything, but gave up and ignored us, not being drawn into the conversation.

Sgt. Green proceeded to outline in great detail how the record of these incidents were erased from official scrutiny, and why.

Radar operators, unlike technicians, were not tested and selected for intelligence, but they naturally had to be very down-to-earth and dependable individuals. Because they had a disproportionately large responsibility for the S.O.B.s (souls on board) the aircraft they were "talking down" to the runway, they were the only "ground-pounders" that I knew of who also received flight pay. I believe this accounted for about a third of their salary, and they also were apparently entitled to many of the other perquisites allocated to flight personnel like better services, etc.

Sgt Green told me when there was a radio conversation regarding a UFO contact among the pilot/flight-crew members, the GCA unit, and/or the Control Tower, and/or Base Operations, that once the incident had ended, by unspoken agreement the reel of two inch wide, slow moving audio tape within the GCA unit which recorded all radio communication would be removed by the operator, bulk erased, and replaced. This is grossly improper behaviour, but even though they were located in my van these magnetic tapes were not part of my responsibility.

When I asked why this was done, I was simply told that that was how it was because the Air Force doesn't believe in flying saucers. I asked what would happen if the operator insisted that he would not break the written rules and continued to insist on the truth. The sergeants acted surprised that I could even imagine such a thing.

Sgt. Green seemed to be very familiar with the procedure, as if it were common knowledge in his speciality. He explained that the first thing that would happen is an immediate removal from regular duty and a loss of flight pay and all associated benefits. Given that most Non-commissioned Officers had families to support and their expenditures equalled their income, that was a powerful deterrent. I kept asking, "But what if he stood his ground?" I was informed that then the operator would be told what he could expect, and might be given a chance to change his mind, but the implication was that the damage would have already been done.

Since UFOs don't exist, the operator could only be hallucinating. A hallucinating radar operator is intolerable, and would minimally be given another career, and certainly an unsavoury and probably very dangerous one. This was at the height of the war in Viet Nam. Naturally, a full-time series of strict psychiatric examinations and medical tests would ensue, and it was made clear to me that all necessary action would be brought to bear in order to satisfactorily resolve the situation. I got the impression that there would be no hesitation to use powerful psychogenic drugs if that was what was necessary. I got the message, loud and clear, and I had been given far more food for thought than I could comfortably digest. I never spoke about what I was told that night to anyone until many years later.

Notes

1. In a previous version of this investigation due to unavailability of the original camera manual at the time, we mistakenly used an incorrect source indicating the longest exposure time for this camera was 1/4 of a second instead of 4 seconds. We correct the error in this book and thank sceptics for pointing it out.

2. For the complete and detailed investigation into the WCUFO's size, refer to Zahi's 74-page report *Analysis of the Wedding Cake UFO; investigations of the WCUFO pictures taken by Billy Meier* or Zahi and Lock's *Researching a Real UFO*.

3. Other investigators found it closer to two metres in diameter.

4. Sphere Reflection Rule c: *No matter what size the reflecting sphere is, the size of the reflected object is always in the same proportion to the sphere's size, given the same distances between the sphere and the object. Alternatively, the absolute magnitude of the reflected image is directly proportional to the sphere's diameter.*

5. "Render" in computer animation jargon refers to a process of creating an image or video as seen from the camera, based on the modelled parameters defined in the computer tool, including among other things, the type of materials and lighting. Animation tools like "Blender" use different "rendering" techniques.

6. We used Image/ Adjustments/ Shadows and Illuminations – increasing shadow values.

7. Deardorff's work on the WCUFO is accessible here: http://www.tjresearch.info/Wedcake.htm. 27 August 2019.

8. The entire video is 49 minutes 39 seconds. Meier's explanation of the narrow field of vision is at the 6-8 minute section, 8 mm movie footage of the pendulum UFO 18:00-21:15 mins.

9. For an explanation of meridians and parallels see http://www.dauntless-soft.com/PRODUCTS/Freebies/Library/books/AK/8-2.htm.

10. See around 3.7 seconds into Langdon's demo at www.youtube.com/ watch?v=QKEf_Xy5Rrc.

11. For a basic introduction on Light Scatter: *Atmospheric light scattering.* http://wiki.flightgear.org/Atmospheric_light_scattering. 31 Aug. 2019. Or see "Scattering of Light": http://www.atmo.arizona.edu/students/courselinks/fall16/atmo170a1s3/1S1P_stuff/scattering_of_light/scattering_of_light.html 6 Oct 2019.

They are Here

Or see "Light Scatter": https://www.itp.unihannover.de/fileadmin/arbeitsgruppen/zawischa/ static_html/ scattering.html. 31 Aug 2019.

[12] For a basic introduction to Mie and Rayleigh light scatter, search Georgia State University: http://hyperphysics.phy-astr.gsu.edu/hbase/atmos/ blusky.html#c3.2 .

[13] Francisco Villate's two Photoshop processed pictures from Frehner's photo #164 (enhanced with Photoshop functions: shadows and Illuminations, semitone, contrast100%, Colour correction-100%, then contrast 100% brightness – 30%).

References

Allfishingbuy.com. "Telescopic fishing poles Hi-quality, powerful, lightweight, sensitive, made of high modulus carbon. Manufactured in Japan." http://www.allfishingbuy.com/Fishing-Pole-14-18.htm#site3. 2 Jan. 2020.

Amazon. https://www.amazon.com/Unconventional-Flying-Objects-Scientific-Analysis/dp/1571740279. 22 Jan 2020.

Blender. www.blender.org. 16 September 2019.

Brookings Institute, The. *Brookings now: Communications, Technology, and Extraterrestrial Life: The Advice Brookings Gave NASA about the Space Program in 1960.* 12 May 2014. "The implications of a discovery of extraterrestrial life." https://www.brookings.edu/blog/brookings-now/2014/05/12/communications-technology-and-extraterrestrial-life-the-advice-brookings-gave-nasa-about-the-space-program-in-1960/ 27 Oct 2019.

Brown, Derren. "Derren Brown Tricks Advertisers With Subliminal Messaging." *YouTube.* 27 July 2016. https://www.bing.com/videos/searchq=darren+brown+subconscious+designs&view=detail&mid=C1EA6D6BE754666B2F0FC1EA6D6BE754666B2F0F&FORM=VIRE.

Butkus, M. http://www.butkus.org/chinon/ricoh/ricoh_singlex_tls/ricoh_singlex_tls.htm. *25 Aug 2019.*

Cambridgecolour. DoF calculator. http://www.cambridgeincolour.com/tutorials/dof-calculator.htm. 26 August 2019.

Cameramanuals. http://www.cameramanuals.org/olympus_pdf/olympus_35ecr.pdf. 25 Aug 2019.

Deardorff, James. *A Refutation of Bruce Maccabee's 1989 Debunking Attempt, and the Episode of the Above-the-treetop Beamship Oscillations.* www.tjresearch.info/ BillyYes.htm

———. *Plausible Deniability.* 2002, updated 2006, 2013. www.tjresearch.info/ denial.htm

———. *Recent Analyses of Meier's Hasenböl Photos Shows their Genuineness.* http://www.tjresearch.info/Hasenbol_Proof.htm

———. *The Wedding-Cake UFOs.* 2006, updated 2013. www.tjresearch.info/Wedcake.htm

Devine, Dyson. Sergeant (Ret.) "*Dyson A. Devine's United States Air Force witness testimony as provided to the Disclosure Project.*" Disclosure Project Witness statement. Private email to the authors dated 24 Jan. 2020.

DOFMaster. www.dofmaster.com/doftable.html. 26 Aug 2019.

Elders, Britt, Lee Elders and Thomas Welch. *UFO Contact From The Pleiades, Volume I.* P 37. Genesis III. 1980. Japanese version p 44. Yaoi, Junichi. Licensee Seishun Best. Tokyo: Onodorisha. 25 June 1992.

Engel, Volker and Marc Weigert. *Uncharted Territory.* 2006.

FIGU org. Beamship – The Metal. Yaoi, Junichi.YouTube. CH-8495 Hinterschimdruti. 16 Oct. 2013. Intercep 1985. Marcel Vogel interview and video analysis of the beamship metal. 23 Feb.2020.

-----------. "Demonstrationsflüge / Demonstration flights (IFOs, not UFOs)." YouTube. http://www.youtube.com/watch?v=EAHfOmvz6_s.

-----------. Hinwil. The two-minute video "The Pendulum UFO." Aufnahme vom. Switzerland: 18 March 1975. *YouTube.* https://youtu.be/K_HnDz4KY6k

-----------. "Pendulum UFO - Demonstrationsflüge / Demonstration flights (detail FIGU)." YouTube. Nov. 2019. https://youtu.be/Gzr3BRUhfy8

Fleming, Dan. www.dofmaster.com/doftable.html. 2005 -2019

FlightGear. *Atmospheric light scattering.* http://wiki.flightgear.org/Atmospheric_light_scattering Sep. 2019.

Frehner, Christian. Private emails to the authors. Fall 2013 – Dec. 2019, March 2020.

------------. Meier's 35 ECR Olympus camera. Video explanation. www.figu.org. https://www.mycloud.ch/s/S0032DBB5C33F3D21EC4CC1B60F5E67B06F405DC88F. 21 Nov. 2019.

FoM. *Future of Mankind.* "Analysis of the wedding cake UFO." http://www.futureofmankind.co.uk/Billy_Meier/Analysis_of_the_Wedding_Cake_UFO#Further_Commentary_by_Dyson_Devine

------------. "Asket and Nera Photos." 2015. www.futureofmankind.co.uk/Billy_Meier/ Asket_and_Nera_Photos. 2017.

------------. "Contact Report 254 Vol. 7." 28 November 1995. http://www.futureofmankind.co.uk/Billy_Meier/Contact_Report_254.

------------. "Contact Report 259 Vol. 7." 25 February 1997. http://www.futureofmankind.co.uk/Billy_Meier/Contact_Report_259 p 460 - 461. 16 Sep. 2019.

------------. "Contact Report 304 Vol. 8." 25 June 2004. http://www.futureofmankind.co.uk/Billy_Meier/Contact_Report_304. Jan. 2020.

------------. "Contact Report 442 Vol. 11." http://www.futureofmankind.co.uk/Billy_Meier/Contact_Report_442. 10 Feb. 2007.

------------. "Contact Report 486 Vol. 12." 11 Jan. 2010. http://www.futureofmankind.co.uk/Billy_Meier/Contact_Report_486 Jan. 2020.

References

-----------. "Photo Gallery." http://www.futureofmankind.co.uk/Billy_Meier/Photo_Gallery. Dec. 2019. Photos #151~#155, #157, #164, #168, #171, #174~ #176, #179, #181. Feb 2020.

Gallery, hd. <www.gallery.hd.org>. Fall of 2013.

Hey Skipper. "*WHY should you use a RIDICULOUSLY Long Fishing Rod?*" www.youtube.com/watch?v=RfBoZtIR0rM).Nov. 2019.

Hill, Paul R. *Unconventional Flying Objects: A Scientific Analysis*. Cover review. Charlottesville: Hampton Roads. 1 December 1995.

Horn, Michael. *The Silent Revolution of Truth*, movie. 2007. www.theyfly.com

------------. *They Fly Blog*. "What Are the Criteria that Reasonable People Use to Determine the Truth?" Dyson Devine's comments near the bottom. https://theyflyblog.com/2014/07/05/criteria-reasonable-people-determine-truth/comment-page-3/#comments. 24 Jan. 2020.

Hyper Physics. Rayleigh and Mie light scatter. Georgia State University. http://hyperphysics.phy-astr.gsu.edu/hbase/atmos/blusky.html#c3.2 Sep. 2019.

Istok, Taro. "Billy Meier Pendulum UFO - Direct Comparison With Suspended Model." 2018. www.youtube.com/watch?v=jKot4oRsvnM

------------. "Billy Meier Pendulum UFO: A Comparison - Meier vs. Langdon (featuring 'Semjase's Pendulum.')" 15 Sep. 2019. https://www.youtube.com/ watch?v=SNlmFfb0ADs.

Lane, Steve. "WCUFO Reflections. View in HD." 10 April 2018. www.youtube.com/watch?v=H7lLE02_WsU

Langdon, Phil. "New! UFO - That's Complete Pendulum! How to reproduce…". www.youtube.com/watch?v=QKEf_Xy5Rrc

Maccabee, Bruce. *Pendulum-like Motion of an Unidentified Object (UO) Filmed by Billy Meier*. http://brumac.mysite.com/Meier/MeierPendulum.htm. 27 Feb. 2020. Previously at http:// www.futureofmankind.co.uk/Billy_Meier/PendulumLike_motion_of_an_unidentified_object_(UO)_filmed_by_Billy_Meier, Maccabee, Bruce, Internet, February 2002.

Michael, Donald N. "The Brookings Report." https://www.brookings.edu/blog/brookings-now/2014/05/12/communications-technology-and-extraterrestrial-life-the-advice-brookings-gave-nasa-about-the-space-program-in-1960 "Proposed Studies on the Implications of Peaceful Space Activities for Human Affairs." p. 182-184. https://www.brookings.edu/wp-content/uploads/2014/05/space_extraterrestrials.pdf. 27 Oct 2019.

Meier, Billy. "The Pendulum UFO" aka "A UFO Circling a Tree." Untitled home movie. Hinwil, Switzerland: 18 Mar. 1975. https://youtu.be/K_HnDz4KY6k.

------------. *Contact Reports Vol. 3*. (*Plejadisch-plejarische Kontakberichte, Gespräche, Block 3*). 375 – 388. "Plejaren Contact Report 123." 4 June 1979.

------------. *Contact Reports Vol. 7. (Plejadisch-plejarische Kontakberichte, Gespräche, 3)*. "Plejaren Contact Report 251." 3 Feb. 1995.

------------. *Contact Reports Vol. 7. (Plejadisch-plejarische Kontakberichte, Gespräche, Block 7)*. 406 – 407. "Plejaren Contact Report 254."28 November 1995.

------------. *Contact Reports Vol. 7. (Plejadisch-plejarische Kontakberichte, Gespräche, Block 7)*. 460 – 461. "Plejaren Contact Report 259."25 Feb. 1997.

------------. *Contact Reports Vol. 11. (Plejadisch-plejarische Kontakberichte, Gespräche, Block 11)*. "Plejaren Contact Report 442." 10 Feb. 2007.

------------. *Through Space and Time.* p 109. Steelmark LLC. 1 Jan 2004.

------------. *Photobuch. Wassermannzeit-verlag,* p 114. Switzerland. 2001.

------------. *Photo-Inventarium,* p 87, 102. FIGU. Augsburg: Wassermannzeit-Verlag, 2014.

Morningstar, Robert. Ed. *UFO Digest* (ufodigest.com) "'The Dancing UFO': -> filmmaker Dan Drasin comments on Rhal Zahi's analysis of a Billy Meier 'Beamship UFO' film." https://www.ufodigest.com/article/the-dancing-ufo-filmmaker-dan-drasin-comments-on-rhal-zahis-analysis-of-a-billy-meier-beamship-ufo-film/. 2 July 2015.

Morrison, Foster. "UFOs – Science and Technology in the Service of Magic." *MUFON UFO Journal.* Number 242, p. 3. June 1988.

Orphancameras. www.orphancameras.com Olympus 35 ECR manual and Ricoh Singlex TLS manual. 26 Aug. 2019.

Paradisopics. <www.papradisopics.com>. Fall 2013.

Parallels. http://www.dauntless-soft.com/PRODUCTS/Freebies/Library/books/AK/8-2.htm

Robinson, Remington. Drawing in BillyMeier's possession. 23 Feb. 2020. www.remingtonrobinson.com

Stevens, Wendelle and Lee and Brit Elders. "Contact – 'Billy' Eduard A. Meier Documentary by Wendelle Stevens (1982)." *YouTube.* www.youtube.com/watch?v=HXLgmXqqrpI

Stevens, Wendelle. C. "Detailed clip of the Pendulum UFO in the 'Contact' documentary." *YouTube.* https://youtu.be/gGCuLIVxxQw

------------.*UFO Contact from the Pleiades – A preliminary investigation report.* 1982. UFO Photo Archives, Tucson, AZ, USA.

------------. *UFO Contact from the Pleiades – A preliminary investigation report.* http://theyfly.com/PDF/PhotoAnalysis.pdf. p. 12. 23 Aug. 2020.

------------. *UFO Contact from the Pleiades – A supplementary investigation report.* 1989. UFO Photo Archives, Tucson, AZ, USA.

References

----------. "UFO Congress, December 1994." *YouTube*. www.youtube.com/watch?v=umJuJ7I_5Go.

Swiss Confederation. "Karten der Schweiz - Schweizerische Eidgenossenschaft", https://map.geo.admin.ch/. 2020.

Szydlik, A. J. "A Technique for calculating atmospheric scattering and attenuation effects of aerial photographic imagery from totally airborne acquired data." Dissertation. School of Photographic Arts and Sciences, Rochester Inst. Technology, New York. 1977. https://scholarworks.rit.edu/cgi/viewcontent.cgi?referer=&httpsredir=1&article=6324&context=theses. 3 Sept. 2019.

Travel British Columbia. Illus. The world's longest fly-rod. http://www.travel-british-columbia.com/north-bc/yellowhead-highway/houston/. 5 Jan. 2014. See also: https://www.roadsideamerica.com/tip/20735. 27 Feb. 2020.

University of Arizona. "Scattering of Light."http://www.atmo.arizona.edu/students/courselinks/spring08/atmo336s1/courses/fall13/atmo170a1s3/1S1P_stuff/scattering_of_light/scattering_of_light.html. 31 Aug. 2019.

Wikipedia. *Atmospheric light scattering*. http://wiki.flightgear.org/Atmospheric_light_scattering. 31 Aug. 2019.

----------. *Interlaced video*. https://en.wikipedia.org/wiki/Interlaced_video

----------. *Al-Aqsa Mosque*. https://en.wikipedia.org/wiki/Al-Aqsa_Mosque

Xiongbin, Chen, Chengyu Min and Junqing Guo. *International Journal of Optics*. "Visible Light Communication System Using Silicon Photocell for Energy Gathering and Data Receiving." Volume 2017, Article ID 6207123, 5 pages. https://doi.org/10.1155/2017/6207123. 12 Aug. 2019.

Yaoi, Junichi. Beamship – The Metal. YouTube. 16 Oct. 2013. Intercep 1985. »Marcel Vogel interview and video analysis of the beamship metal. 23 Feb. 2020.

----------. "FULL Billy Meier-1985 Beamship - The Movie Footage." Nippon Television. YouTube. Meier interview. David Seaton. Circa 1978. Interview at 1:50 mins. Camera demo at 4:50 mins, movie camera demo at 7:20 mins. Pendulum UFO 8 mm movie footage at 18:00-21:15 mins. www.youtube.com/ watch?v=K58MjoKSMPo. YouTube https://youtu.be/WkQgw1PPLZM unavailable.

----------. "The 'Billy Meier contact case' in Japanese (Nippon TV)." Figu copyright. 9 Oct. 2019. YouTube. https://www.youtube.com/watch?v=xvbtGJKljQg.

----------. *UFO···Contact From The Pleiades. (UFO puleadesu hoshidan kara no sekkin.* プレアデス星団からの接近.) P 37 & 44. Japanese trans from U.S. version Elders, Britt, Lee Elders and Thomas Welch. *UFO*

Contact From The Pleiades, Volume I. version. Licensee Seishun Best. Genesis III. 1980. Tokyo: Onodorisha. 25 June 1992.

Zahi, Rhal. *Analysis of the Wedding Cake UFO; investigations of the WCUFO pictures taken by Billy Meier.* PDF file 4.0MB. Mar. 2013. http://www.rhalzahi.com/docs/WCUFO-EN-v2.pdf. Also: http://www.futureofmankind.co.uk/Billy_Meier/Analysis_of_the_Wedding_Cake_UFO. 6 Oct. 2013.

----------. *An Investigation Into the Pendulum UFO – An Analysis of a Billy Meier Film.* PDF file 6.3MB. Jan. 2014, Ver. 5. http://www.rhalzahi.com/docs/pendulum-EN.pdf.

----------. A real ET space ship? (WCUFO). Jan. 2013. *YouTube.* www.youtube.com/watch?v=6WHqBvBZOqg

----------. New revelations of an UFO – WCUFO. Feb. 2014. *YouTube.* www.youtube.com/watch?v=_pjcbF1oK8Q

----------. OVNI Danzante – Dancing UFO. 29 June 2015. *YouTube.* www.youtube.com/watch?v=IKeutVKFbG0

----------. The Pendulum UFO – Jumping in the Space. Jan. 2014. *YouTube.* www.youtube.com/watch?v=8BI-pt3L9I4

----------. Worldwide UFO Controversy. June 2018. *YouTube.* https://www.youtube.com/watch?v=YsPK_XvfqeI&t=1863s

----------. Yes… it is a real UFO (WCUFO). Aug. 2013. *YouTube.* www.youtube.com/watch?v=JoZKwqptZ2Y

Zahi, Rhal and Christopher Lock HonFSAI. *Researching a Real UFO: A Practical Guide to WCUFO Experimentation for Young Scientists.* Create Space (Amazon.com). Nov. 2016.

Zawischa, Dietrich. "Light Scatter." https://www.itp.unihannover.de/fileadmin/arbeitsgruppen/zawischa/static_html/scattering.html. 31 Aug. 2019.

www.ingramcontent.com/pod-product-compliance
Lightning Source LLC
Chambersburg PA
CBHW061128170426